The Bright Dark Ages

History of Science and Medicine Library

VOLUME 53

Knowledge Infrastructure and Knowledge Economy

Edited by

Karel Davids (*VU University, Amsterdam*)
Larry Stewart (*University of Saskatchewan, Saskatoon*)

VOLUME 5

The titles published in this series are listed at *brill.com/hsml*

The Bright Dark Ages

Comparative and Connective Perspectives

Edited by

Arun Bala and Prasenjit Duara

BRILL

LEIDEN | BOSTON

Want or need Open Access? Brill Open offers you the choice to make your research freely accessible online in exchange for a publication charge. Review your various options on brill.com/brill-open.

Typeface for the Latin, Greek, and Cyrillic scripts: "Brill". See and download: brill.com/brill-typeface.

ISSN 1872-0684
ISBN 978-90-04-26418-2 (hardback)
ISBN 978-90-04-26419-9 (e-book)

Printed by Printforce, the Netherlands

Contents

Acknowledgements

The chapters in this volume developed from papers presented at two conferences, *The Bright Dark Ages: Rethinking Needham's Grand Question* and *The Bright Dark Ages: Comparative and Connective Perspectives*, held in Singapore in 2010 and 2013. We would like to thank the sponsorship of the Institute of Southeast Asian Studies (Singapore), the National University of Singapore and Nanyang Technological University for the first conference, and Asia Research Institute (National University of Singapore), Situating Science Strategic Knowledge Cluster (Canada) and the University of Iceland for the second. Our gratitude also goes to the editors of the Brill Series *Knowledge Infrastructure and Knowledge Economy*, Karel Davids and Larry Stewart, for their unrelenting critiques and insightful comments. Finally we would like to recognise the initial collaboration of Monique van Donzel, and the support of our Brill editors Sabine Steenbeek, Michiel Thijssen, Wendel Scholma and Thalien Colenbrander.

List of Contributors

Arun Bala

is a physicist and philosopher of science who is Senior Research Fellow with the Asia Research Institute, National University of Singapore, and currently Visiting Professor, Department of Philosophy, University of Toronto. His publications include *The Dialogue of Civilizations in the Birth of Modern Science* (Palgrave, Macmillan, 2006), and *Asia, Europe and The Emergence of Modern Science: Knowledge Crossing Boundaries* (ed., Palgrave Macmillan, 2012). His recent research focuses on the implications for philosophy of science of the dialogical connections across the Eurasian region that shaped the rise of contemporary scientific ideas in the early modern era.

Andrew Brennan

is currently Professor of philosophy and Pro Vice-Chancellor at La Trobe University, Melbourne, Australia. From 1992 to 2006, he was professor and chair of philosophy at the University of Western Australia. His recent books include two co-authored works – *Understanding Environmental Philosophy* (Acumen, 2010), *Logic: Key Concepts* (Continuum, 2005) – and also *Thinking About Nature* (Routledge 2014). As well as publications in the history of science, he has also written extensively on environmental ethics, public policy and philosophy of logic.

James Robert Brown

is a philosopher of science at the University of Toronto. His interests include: scientific realism, thought experiments, philosophy of mathematics, foundations of physics, and the social relations of science. More recently he has developed an interest in the relations between modern science and local or traditional knowledge. His published books include: *Who Rules in Science? An Opinionated Guide to the Wars* (Harvard, 2001), *Philosophy of Mathematics: An Introduction to the World of Proofs and Pictures* (Routledge, 2nd edition 2008), and *The Laboratory of the Mind: Thought Experiments in the Natural Sciences* (Routledge, 2nd edition 2010), and most recently, *Platonism, Naturalism, and Mathematical Knowledge* (Routledge, 2012).

Prasenjit Duara

is the Oscar Tang Chair of East Asian Studies at Duke University. He was born and educated in India; received his PhD in Chinese history from Harvard University. He was previously Professor and Chair of the Dept of History and

Chair of the Committee on Chinese Studies at the University of Chicago (1991–2008). He was also Raffles Professor of Humanities and Director, Asia Research Institute at National University of Singapore (2008–2015).

George Gheverghese Joseph

is Honorary Reader in the School of Education, University of Manchester. His recent publications include five books: *Women at Work* (Philip Allan, Oxford, 1983), *Multicultural Mathematics: Teaching Mathematics from a Global Perspective* (Oxford University Press, 1993), *George Joseph: Life and Times of a Kerala Christian Nationalist* (Orient Longman, 2003), *A Passage to Infinity: Medieval Indian Mathematics from Kerala and its Impact* (Sage Publications, 2009) and *The Crest of the Peacock: Non-European Roots of Mathematics* (Princeton University Press, 2000, 2011).

Henrik Lagerlund

received his PhD from Uppsala University in 1999 and is currently Professor and Chair of the Department of Philosophy at the University of Western Ontario. He has published extensively on medieval philosophy, including the monograph *Modal Syllogistics in the Middle Ages* (Brill, 2000). Among his edited books are *The Philosophy of Francisco Suarez* (Oxford, 2012), *Rethinking the History of Skepticism* (Brill, 2010), and *Representation and Objects of Thought in Medieval Philosophy* (Ashgate, 2008). He is also the editor-in-chief of the *Encyclopedia of Medieval Philosophy* (Springer, 2011). His current research project is on the late medieval background to the mechanical philosophy in seventeenth century Western philosophical tradition.

Norva Y.S. Lo (勞若詩)

is currently a senior lecturer in Philosophy at La Trobe University. She researches in the areas of moral philosophy, environmental philosophy, animal ethics, the philosophy of David Hume, and also teaches logic. Some recent publications include *Understanding Environmental Philosophy* (Acumen, 2010), and chapters in *Hume on Motivation and Virtue* (Palgrave Macmillan, 2009), *The Routledge Companion to Ethics* (Routledge, 2010), *Perspectives on Human Suffering* (Springer, 2012), and *The Routledge Handbook of Global Ethics* (Routledge, 2015). She is also the co-author of the "Environmental Ethics" entry in the *Stanford Encyclopedia of Philosophy*.

Roddam Narasimha

currently DST Professor at the Jawaharlal Nehru Centre for Advanced Scientific Research, was formerly Director of the National Aerospace Laboratories and the National Institute of Advanced Studies, all at Bangalore. Much of his professional career has been spent at the Indian Institute of Science, also at Bangalore. He was educated there and at the California Institute of Technology, where he obtained his Ph.D. His scientific work has been largely in the subject of fluid dynamics, and in particular its applications in aerospace technology and atmospheric sciences. His interests in the history of science are particularly concerned about the epistemology underlying classical Indic thinking about science and scientific method. In particular he has proposed that the epistemology underlying Indian mathematical astronomy is best thought of as computational positivism, an approach that helps in understanding Indic ideas about proofs and pramanas. He has written several papers on this and related subjects, is the co-editor (with Helaine Selin) of *The Encyclopaedia of Classical Indian Sciences* (Universities Press, 2007) and co-editor of *Nature and Culture* (Munshiram Manoharlal, 2011). He has been elected to the Royal Society, the US National Academy of Science, the US National Academy of Engineering and the American Academy of Arts and Sciences, and won the Trieste Science Prize in 2009. In 2013 he was awarded the second highest civilian honour in India, Padma Vibhushan, by the President of the Republic.

Hyunhee Park
PhD Yale, Associate Professor of History at the City University of New York, John Jay College, specializes in the cross-cultural contacts of premodern Afro-Eurasia and teaches Chinese history and global history. Her book *Mapping the Chinese and Islamic Worlds: Cross-Cultural Exchange in Pre-Modern Asia* (Cambridge University Press, 2012) explores medieval contact and exchange between the Islamic World and China by utilizing geographic and cartographic information. Her new research project encompasses world mapping and other types of information transfers spanning medieval Afro-Eurasia and the early modern Atlantic World. She is currently serving as an assistant editor of the academic journal *Crossroads – Studies on the History of Exchange Relations in the East Asian World*.

Franklin Perkins
is Professor of Philosophy at DePaul University (Chicago) and Associate Professor of Philosophy at Nanyang Technological University (Singapore). His main research is on Classical Chinese Philosophy, Early Modern European Philosophy, and Comparative Philosophy. He is the author of *Leibniz and China: A Commerce of Light* (Cambridge University Press, 2004), *Leibniz: A Guide for the Perplexed* (Continuum, 2007), and *Heaven and Earth are not*

Humane: The Problem of Evil in Classical Chinese Philosophy (Indiana University Press, 2014), and he was co-editor of *Chinese Philosophy in Early Excavated Bamboo Texts* (*Journal of Chinese Philosophy Supplement* 2010) (with Chung-ying Cheng), and *Chinese Metaphysics and Its Problems* (Cambridge University Press, 2015) (with Chenyang Li).

Hans Pols

is associate professor at the Unit for History and Philosophy of Science at the University of Sydney. He is interested in the history medicine, in particular the history of psychiatry. He has conducted research into the history of the mental hygiene movement, the treatment for nervous breakdown in the armed forces during World War II, and on the history of medicine in the former Dutch East Indies and modern Indonesia.

Kapil Raj

is Directeur d'études (research professor) at the École des Hautes Études en Sciences Sociales in Paris and Dean of doctoral studies in the history and sociology of science. His research questions the commonly held assumption of the western origins of modern science. His book, *Relocating Modern Science* (Palgrave, 2007) focuses on the role of circulation and encounter between South Asian and European skills and knowledges in the emergence of crucial parts of modern science. He also co-edited *The Brokered World: Go-Betweens and Global Intelligence, 1770–1820* (Science History Publication, 2009) and has just completed co-editing a collective work on the history of knowledge and science in the long 19th century scheduled to appear in French later in 2015. He is currently engaged in researching for his next book in the urban and knowledge dynamics of Calcutta in the 18th century.

Sundar Sarukkai

is currently at the Manipal Centre for Philosophy & Humanities, Manipal University where he was the Founder-Director from 2010–2015. He was earlier at the National Institute of Advanced Studies, Bangalore from 1994 to 2009. He is the author of *Translating the World: Science and Language; Philosophy of Symmetry* and *Indian Philosophy and Philosophy of Science*. Two of his recent books are *What is Science?* (National Book Trust, 2014) and *The Cracked Mirror: An Indian Debate on Experience and Theory* (Oxford University Press, 2012; co-authored with Gopal Guru). He is an Editorial Board member of the *Leonardo Book Series* published by MIT Press on science and art, as well as co-editor of three volumes on logic. He is an Editorial Advisory Board member of the

Leonardo Book Series on science and art, published by MIT Press, the Series Editor for *Science and Technology Studies*, Routledge and the Chief Editor of the Springer *Handbook on Logical Thought in India*.

Mohd. Hazim Shah

holds a Bachelor's degree in Liberal Studies in Science from Manchester University, England, a Master's degree in Philosophy from the London School of Economics, and a PhD in the History and Philosophy of Science from the University of Pittsburgh, U.S.A. In 1993 he was a Visiting Research Fellow at the Department of History & Philosophy of Science, University of Melbourne, Australia, and spent time as a Visiting Scholar at the Department of History and Philosophy of Science, University of Cambridge, in 2008. He was formerly a Professor at the Department of Science & Technology Studies, University of Malaya, where he served as Head of Department from April 2001 to August 2007. He currently serves as a Senior Research Fellow in the same department after his retirement in May 2015. He was the President of the Malaysian Social Science Association from 2010 until 2014. His research interests include theoretical studies on science and culture, and comparative epistemology.

Geir Sigurðsson

is Associate Professor of Chinese Studies and currently Head of Faculty of Foreign Languages, Literature and Linguistics, School of Humanities, University of Iceland. He received his Ph.D in Philosophy from the University of Hawai'i. His areas of specialization include Chinese philosophy, Chinese studies, philosophy of education, ethics and continental philosophy. He is author of *Confucian Propriety and Ritual Learning: A Philosophical Interpretation*, published by SUNY Press in 2015.

Cecilia Wee

is Associate Professor at the National University of Singapore. She received her Ph.D in Philosophy from the University of Pittsburgh. Her main area of research is in Descartes and early modern philosophy. She also has some interest in environmental ethics and comparative philosophy, and is currently reviving a (much) earlier interest in the philosophy and history of science. She has published a book, *Material Falsity and Error in Descartes' Meditations* (Routledge, 2006). Her papers have appeared in journals such as the Canadian Journal of Philosophy, British Journal for the History of Philosophy, Environmental Ethics and History of Philosophy Quarterly, as well as various edited volumes.

[handwritten margin notes: "Earlier / Hellenistic / golden age"]

[handwritten: "6 BCE — 9 BCE / European dark ages / Asia blooming / (bright dark ages)"]

[handwritten: "Yeh Yeh, Science but / but not The real"]

Introduction

Arun Bala and Prasenjit Duara

[handwritten: "A) Comparative + Needham"]

[handwritten: "B) Connective, sustained interaction"]

Attempts to globalize history, philosophy and sociology of science by including the content and contexts of the Asian traditions – especially Chinese, Indian and Arabic-Islamic – along with Western/modern science have generally adopted two different approaches. Both are inspired by the recognition that while the period between the sixth and sixteenth centuries constitutes the dark age of science in Europe in contrast to the golden age of Hellenistic science, science in Asia was blooming, with seminal advances being made in China, India and the Arabic-Islamic world – a period we shall refer to as the "Bright Dark Ages". However, these approaches have proceeded in opposing directions. One looks at the Asian and Western/modern traditions in a comparative perspective largely inspired by an attempt to answer the famous Needham Question: Why did modern science emerge in Europe but not Asia despite the great achievements of Asian sciences in the Bright Dark Ages? The second involves a connective approach, showing how modern science became possible only through a sustained interaction between practitioners of specialized knowledge cultures from various parts of Europe and Asia – and indeed from other parts of the world.[1]

Although the two approaches are not necessarily incompatible they often originate from profoundly different conceptions of the history of science with corresponding implications for our understanding of both the philosophy and sociology of science. Comparativists generally assume that the traditions of science in different civilizations grew in isolation from each other, and that connections across civilizations had only a marginal impact on the science that developed within civilizations (with the possible exception of Arabic-Islamic science that acknowledges contributions from other civilizations as crucial).[2] This has implications for how we comprehend the philosophy and

---

1 For a comprehensive comparative perspective see Ronan (1983). More recently connected perspectives have also become significant. See Joseph (2011), Hobson (2004), Bala (2006) and Raj (2007).

2 This is evident in the seminal historical text by al-Andalusi (1991) which discusses the contributions of various cultures – Egyptian, Arab, Greek, Indian, Chaldean, Chinese among others – in promoting the growth of scientific knowledge.

   Of course essentialized notions of civilization and science are both highly contested and questionable. However, this can be avoided if we take civilization to be an open-ended and evolving sphere of inter-referential ideas and practices. This sphere, nonetheless, can draw from – often unacknowledged – sources outside the sphere. Similarly, a non-essentialized

sociology of science in each of these civilizations – it suggests that the crucial philosophical and socio-cultural impacts must be sought for in essential factors within the civilization concerned rather than outside it.

By contrast connectivists not only perceive the Asian traditions as influencing each other crucially in the Bright Dark Ages, but also see these traditions as shaping the emergence of modern science in Europe. Such a connective history of science also impacts on our conception of the philosophical ideas and social contexts that shaped a particular scientific tradition. Each tradition within a particular civilization is now seen as growing through an exchange of philosophical ideas with other traditions in a socio-cultural context defined by other civilizations.[3]

Such a perspective appears novel in the modern era because the prevalence of Eurocentrism in the history of science in recent times has not only obscured the contributions of Asian, and other non-European, knowledge to modern science, but also treated Asian traditions of science as growing within insular Chinese, Indian or Arabic enclaves. The Arabic tradition is even largely seen as merely preserving the Greek heritage for transmission back to the West. But the view in the Bright Dark Ages was quite different. Writing in Europe in Andalusia in the 11th century Said al-Andalusi (1029–1070), in his renowned history of science *Categories of Nations* describes how Persian, Chaldean, Turk, Egyptian, Greco-Roman and Indian civilizations had made contributions to advance scientific knowledge.[4] Similarly his fellow contemporary in the East in India, Abu Rayhan al-Biruni (973–1048), compared and contrasted Indian, Greek, Arabic and other traditions of science in order to understand how they can be made to further contribute to the shared pool of scientific knowledge.[5]

Such an inclusive attitude to science was not peculiar to the Arabs alone. In the sixth century Sanskrit encyclopaedia, *Brihat-Samhita*, the Indian mathematical astronomer, Varahamihira (505–587 CE) writes: "The Greeks, though foreign, must be honoured since they were trained in Sciences and therein,

---

conception of science would reject strict epistemological criteria such as falsifiability proposed by Popper, and Kuhn's sociological criterion which leads him to maintain that no other place outside Europe "supported the very special communities from which scientific productivity comes." Kuhn (1970), pp. 167–8. But if we reject such specialized notions of science, and treat it as any systematized body of natural knowledge, it gives us good reason to assume that scientific knowledge has been produced by many cultures. For a more detailed discussion of these matters see Bala & Joseph (2007).

3   Recent trends in global history have begun to move in this direction. For a comprehensive survey of such approaches see Duara et al. (2014).

4   Strangely al-Andalusi (1991, p. 7) concludes that the Chinese showed no interest in science although they surpassed other nations in the industrial and graphic arts.

5   See al-Andalusi, Said (1991) and al-Biruni (2005).

excelled others."[6] Varahamihira's major work *Pancha-Siddhantika* (Treatise on the Five Astronomical Canons) summarises five earlier astronomical works, two of which, *Romaka Siddhanta* and *Paulisa Siddhanta*, refer to works based on the learning of Byzantine Roman and Alexandrian Hellenistic astronomical traditions respectively. Hence, Indian astronomers were well aware of the significant contributions made by the Greek tradition to the advancement of scientific ideas, especially mathematical and astronomical, in the Bright Dark Ages.[7]

The Chinese, too, had a long tradition of inviting non-Chinese astronomers into their Astronomical Bureau – the Indians in the Tang dynasty (618–907) at the time of significant Buddhist influence in China and the Imperial Court, the Muslims in the Mongol Yuan dynasty (1271–1368) when contacts with West Asian astronomers were established through the corridors of communication across the steppes, and the Jesuits in the Ming dynasty (1368–1644) when communications were established between Europe and China across the oceans. The ecumenicity of the sources of scientific knowledge was thus far more acknowledged in China, India and the Arabic-Islamic world during the Bright Dark Ages than it has been in Europe until recent times.[8]

The study of the growth of scientific knowledge during the Bright Dark Ages also has become more imperative after Joseph Needham's pioneering studies documented the numerous contributions of Chinese science to modern science. His work has inspired a vast body of literature which has also come to document the contributions of other Asian traditions of science, especially the Islamic-Arabic and Indian, to modern science. Hence, the history of science in the Bright Dark Ages can no longer be considered to be sundered from the history of modern science. This paves the way for possible comparative studies between the circulation of science across cultures in the Bright Dark Ages, and the circulation of science from the Bright Dark Ages into modern science.

In recent years expanding interest in globalizing the historical roots of modernity has also led to a widening interest in globalizing the history of science. Consequently there has been an increase in the array of publications involving exchanges of scientific knowledge in areas as diverse as mathematics, astronomy, botany, *materia medica* and geography among others. Although studies of this genre have focused on history they also incidentally raise problems in the social epistemology of science that are extremely relevant for

---

6   Varahamihira (2011), *Brihat-Samhita* 2.15. Also quoted in al-Biruni (2005), p. 23.

7   For a more detailed discussion see McEvilley (2001) and Sarma (2000).

8   Many studies have made this point including Ohashi (2008), Rufus (1939), Needham with Wang Ling (1959) and Elman (2009).

understanding more deeply comparative and connective issues in science. This is because even acknowledging that the histories of science across Eurasia are connected does not erase the need for comparative exercises. In many ways it even accentuates the need for them. For instance, in a recent work, George Steinmetz utilizes critical realism to grasp how comparative histories can involve an understanding of deeper connections at the level of generative causal mechanisms or a conjunction of multiple causal mechanisms that may not necessarily be expressed at the surface level.[9] This kind of conjoined methodology may be able to address Needham's question more usefully than it has so far. Thus certain critical influences, say in astronomy or mathematics or experimentalism from one or more cultures may be present in the practical or technological culture of another society. Yet when the relevant theory is developed in the second society, that particular idea may be folded into another basic principle where it may not be easily identifiable.[10]

Despite his acknowledgment of the historical contributions of Asian civilizations to modern science, Needham nevertheless posed his famous question "Why did modern science emerge in Europe and not elsewhere?" Moreover, although Needham's main focus was on documenting Chinese contributions, he never failed to point to the need to take into account and unearth for history the Indian and Islamic contributions to modern science. Needham himself recommended such a reorientation when he wrote:

> It is necessary to see Europe from the outside – to see European history and European failure no less than European achievement, through the eyes of that large part of humanity, the peoples of Asia and indeed also of Africa. (Needham 1979: 11)

Again,

> I have pictured modern science as being like an ocean into which the rivers of all the world's civilizations have poured their waters. (Needham 2004: 201)

---

9   Steinmetz in Duara et al. (2014).

10   Some of the papers in this volume serve to illustrate this – especially Narasimha and Joseph in mathematics, and Wee in experimentation. For astronomy see Bala (2006). The study by Hobson (2004) give a systematic excavation of such enfoldments that have shaped the modern natural and social sciences.

Nevertheless, despite his commitment to a connective, entangled, dialogical and circulatory history of modern science transcending the Eurocentric historical approaches which preceded him, Needham remained fundamentally Eurocentric in his epistemological and sociological orientation. This is evident in the reasons he gives for why modern science did not develop in China. On the one hand he blames China's bureaucratic feudal culture quite different from Europe's aristocratic feudalism which he thinks obstructed the emergence of mercantile and industrial capitalism. Secondly modern science depended on a mathematical rationality built upon Greek tradition – especially Euclidean geometry – which China also lacked. Thus Chinese science could not go beyond the ethnic bound categories that allowed the exchange of technical discoveries but not theoretical ideas. This would become possible only with mathematical reasoning and hypotheses which, being ecumenical, could be assimilated by all human cultures. Thus we can see Needham as offering a connective history of science, but a comparative sociology and epistemology of Chinese and Western science. In effect he sets in motion two trends – a comparative and connectionist – that continue to engage each other in uneasy dialogue.[11]

The contributors in this volume may be seen as broadly addressing issues in the philosophy, history and sociology of science that take into account comparative and connective concerns generally ignored when the impact of Eastern traditions of science in understanding modern science is overlooked. Many of the papers also implicitly or explicitly question the uneasy narrative between connected histories and separate socio-epistemologies that Needham employed. Taken together they require us to rethink how we should approach epistemological issues associated with the discovery and justification of knowledge, historical processes connected with the circulation of ideas across Asian traditions of science that promoted the growth of knowledge, and the social contexts that shaped the development of science and technology in both the Bright Dark Ages and the early modern era.

There are many epistemological questions that arise when traditions of knowledge are seen to both collide and fuse with each other, such as: Should traditional forms of knowledge be seen as *praxis* and *techne* rather than as ways of understanding and theorizing the world? Is it really legitimate to focus on indigenous technological knowledge as a resource for science while ignoring the theoretical perspectives that inform these traditional knowledge systems? Would this imply that science is not a universal system of knowledge?

11   Joseph Needham, 'The Roles of Europe and China in the Evolution of "Ecumenical Science,"' in Needham (1970) p. 397.

What are the implications for scientific method if ideas from a plurality of traditions of science contributed to modern science? Particularly significant are the attempts by the writers to investigate what theoretical and philosophical resources are available in the various Asian traditions that could enrich our understanding of modern science and its philosophy.

Interwoven with the epistemological concerns are also historical questions. There is today, a vast literature that addresses the circulation of species, peoples, goods, technologies, ideas and practices across Eurasia. Jack Goody, David Pingree and others have repeatedly shown that Eurasia has been basically unified since the onset of the Bronze Age. Pingree, a leading historian of the exact sciences, has declared, as "a simple historical fact, scientific ideas have been transmitted for millennia from culture to culture, and transformed by each recipient culture into something new." (Pingree 1992: 563) The early technological revolution enabled intensive agriculture that permitted surplus accumulation. Indeed, class stratification, literacy, bureaucracy, urbanization and inter-linked city-based civilizations across the Eurasian zone was made possible by the technological revolution that enabled intensive agriculture. Consequent upon these factors and depending on the changing circumstances of access in a particular region, the zone came to be connected by long distance trade, warfare, technology, ideas and organizational and institutional practices (including mass industrial production). (Hellenthal et al. 2014)

In some of his recent works, the Cambridge anthropologist Jack Goody argues that European scholars have striven to disembed 'Western' history from its actual, interactive historical context starting from Antiquity to the present. Carthage, Phoenician traders, Persia, Egypt, India, China and the Islamic world played a much greater role in Western histories than even the most radical analysts have acknowledged. Thus while the industrial revolution of the late 18th century was undoubtedly a northern European achievement, this revolution cannot be understood without the background of "industrialization, mechanization and mass-production developed...in China with textiles, ceramics, and paper, in India with cotton, later taken up in Europe and the Near East..." Goody also discusses the realm of knowledge production and values, including not only scientific subjects such as mathematics, medicine, and astronomy, but also humanism, democracy, individualism and even romantic love. He does not find European societies to be the exclusive proprietors of these institutions and values in any of these cases. Rather the Europeans owed these traditions in a significant way to other, most notably, Islamic societies. (Goody 2006: 210)

At different times, different cities or regions in Eurasia – whether joined by trading, religious and technological networks or separated by empires, disease, political instability, piracy or climate change – created different nodes of absorption, innovation or isolation in this gradually expanding zone over the millennia. Seen in this wider canvas of changing zones of innovation, the capacity of some European social formations to institutionalize secular knowledge production and sustained economic growth over the last two centuries represents a very important but not unique achievement. Other societies were poised to learn the lessons from one part of the Eurasian world quickly enough – given the time-scale of history – and it may be unproductive to isolate a region or part of the world as having contained within the seeds and sources of science since the beginning of time.

This raises the questions: To what extent did commerce drive the circulation of scientific knowledge across civilizations in the Bright Dark Ages? What role did inter-civilizational dialogue play in promoting or hindering scientific and technological capabilities in the Bright Dark Ages? How did knowledge from the Bright Dark Ages come to be transformed and assimilated into modern science? Did mathematics and medicine, and their transmissions across the various Asian civilizations, constitute a significant component that contributed to advancing knowledge in the Bright Dark Ages?

Furthermore the epistemological and historical issues cannot be separated from the sociocultural contexts that shaped the growth of scientific knowledge in the various Asian civilizations during the Bright Dark Ages. Particularly significant are the pan-Eurasian trading networks both by land and sea that linked China, Southeast Asia, India, the Middle East and Europe. These raise questions such as: What sociocultural contexts hinder or promote scientific knowledge through intercultural exchanges? Do such factors determine whether the processes of transfer occur through diffusion or appropriation?

The manner in which philosophical, historical and socio-cultural aspects fold and get entangled with each other makes it difficult to separate the chapters into sharply delineated sections that focus on a single aspect. Hence the contributions have not been divided into neat sections although broadly speaking the focus in the earlier chapters tend to be more epistemological and philosophical, and that of the later chapters more historical and sociological.

This volume commences with Andrew Brennan's and Norva Lo's essay which raises the question: If scientific traditions evolve, then what are the historic counterparts to the central components in a biological evolutionary process, namely variation, multiplication and heredity? They maintain that in the competition among scientific theories, as with worldviews, religions and ideologies, there will be a variety of discourses that struggle for existence through competition with each other in the attempt to pass on their characteristics to their successors. It leads them to conclude that viewed through the lens

of evolutionary theory, geographical and cultural isolation can be an explanatory factor for the development not only of stable biological lines, but also of
theoretical lineages (such as some of the "traditions" in the history of science).
By contrast, mixing of cultures at times of social upheaval, war, colonization
and large population movements can encourage the dialogue and debate that
drive the emergence of new lines of inquiry (analogous to periods of rapid
speciation, or to periods of borrowing between different languages). Brennan
and Lo use this analogy with biological speciation and the recognition of the
part played by groupings or constellations of thinkers, to evaluate the methodologies of comparative and connectivist histories of science. By plotting the
descent of the modern sciences by means of a "theory tree" they show that
we can understand the influences of earlier theories on later ones (as argued
by connectivists), while also making space for cases of parallel evolution (as
assumed by comparativists). They conclude that although models of linguistic
and biological evolution fail to provide a perfect analogy for theoretical evolution they challenge Needham's metaphor of older streams of science flowing
into the ecumenical ocean of modern science.

     While Brennan and Lo's chapter looks at the parallels between conceptual
and biological evolution to challenge what they see as Needham's Whiggish
conception of modern science as the culmination of all other traditions of
science, Sundar Sarukkai questions the equally Whiggish notion that only
Western philosophy can be used to understand modern science. He notes that
there is no doubt now that seminal contributions in the fields of science, technology and mathematics from Asian cultures influenced the development of
modern science in Europe. Yet these historical records do not seem to have
made any dent on the community of philosophy of science which continues
to draw its philosophical resources from early Greek thought onwards but has
ignored any possible engagement with non-Greek and non-European philosophies. He finds this surprising particularly because many traditions of Indian
philosophy were not only engaged with the development of rational structures
of knowing, including logic, but some of them also developed sophisticated
theories of realism, which are potentially useful in the debates concerning
scientific realism. Hence Sarukkai's chapter attempts to extract philosophical themes that are catalyzed by connective histories of science by specifically
looking at selected Indian mathematical and scientific texts, and the ways
in which they conceptualized notions of matter, logic and mathematics. He
argues that by paying attention to alternative traditions of scientific practices
and methods and the ways in which they contributed to science we may even
be led to a 'different' philosophy of science – one that does not impugn the

notion of a cosmopolitan science although it grew by drawing on many local traditions of science.

James Robert Brown develops a similar effort to look at epistemological debates in science through different cultural frameworks. He does this by drawing on the literature on Galileo's thought experiments to answer Needham's Grand Question: Why was China surpassed scientifically by the West?, or, to put it another way, Why did Europe, unlike China, have a scientific revolution, given that China was so well prepared for it? Although this question has prompted many different answers and extensive, on-going scholarly debate, Brown argues that historiographic and philosophical work on Western science and its past has raised several conceptual problems that might carry over to comparative studies. He notes that Thomas Kuhn, for instance, claims that people with different paradigms, even in a single culture, "live in different worlds" and consequently talk past one another, and asks what implications this might have when we try to compare Western science with Chinese science. He considers this profoundly relevant for the Needham question since Richard Nisbett and his co-worker Kaiping Peng use responses to Galileo's thought experiment to claim that Easterners and Westerners think in fundamentally different ways: the former are dialectical and holistic, while the latter are analytic and reductionist. Although Nisbett and Peng treat these results as providing an answer to the Needham question, since early modern science is rooted in analytic thought, Brown contests this conclusion by looking more closely at their experimental findings in the light of Kuhn's incommensurability thesis.

In the following chapter Geir Sigurðsson also turns to the contrast between holistic and analytical thinking that separates classical Chinese thinking from modern European thought. But he adds a twist to it by questioning the pervasive assumption made by the vast majority of Euro-American and Asian intellectuals, at least since Hegel and until recently, that Chinese philosophy, having failed to produce anything comparable to the Scientific Revolution in Europe, has little to offer of scientific (and even philosophical) value. He suggests that the relatively recent wave of scholarly interest in classical Chinese thinking requires us to revise this time-honored assessment. He admits that scholarly advocates for the contemporary value of Chinese philosophy tend to focus on social and ethical implications of its teachings, and, to the extent that they associate them with modern science, on their meta-philosophical roles as humanist or eco-friendly worldviews restricting the scope of science and technology. Nevertheless he notes that some have ventured further than this and argued for a meaningful parallelism between Daoist, Neo-Confucian, and Buddhist ontologies, and an emergent worldview proceeding from contemporary physics, phenomenology and post-structuralist suggestions. Although this, in turn, has prompted a critical response by both scholars of science, philosophy and sinology, Sigurðsson seeks to make sense of recent positive assessments of the Chinese philosophical worldview by discussing some underlying key-notions that show affinities between the world view implied by quantum mechanics and classical Daoist and Neo-Confucian traditions. It suggests that Chinese traditions could provide a vocabulary to describe the cosmological implications of quantum physics and be better placed to understand the status and role of science and technology in the contemporary world.

In contrast to the earlier chapters which focused on either Indian or Chinese traditions of science, Cecilia Wee addresses what she considers to be a major contribution of Arabic science. Specifically she argues that Ibn al-Haytham (Alhazen) is justly famed, not just for his contribution to optics, but for his role in developing the methodology of the experiment. She seeks to evaluate al-Haytham's contributions to the development of the scientific experiment, now employed not just in the West but in nearly every country in which scientific research takes place. She maintains that the contribution of al-Haytham is greater than some strands in recent philosophy of science allow. She notes that since the publication of Kuhn's *Structure of Scientific Revolutions* there has developed a strong line of argument that experimental observations are not 'objective' – that they are always informed and shaped by the theoretical commitments of the observer. Although this somewhat de-values the scientific experiment, which had previously been seen as the arbiter of the acceptability of a particular scientific theory, she argues that the phenomena observed in experimental contexts are often robust enough to withstand the apparent theory-dependence of observation, and hence can play an independent and significant role in the evaluation of theories. She concludes that considered in this light, al-Haytham's development of the scientific experiment during the Bright Dark Ages retains its place as a crucial contribution to the overall development of the scientific method.

As with Cecilia Wee, Henrik Lagerlund also points to Arabic influences on modern science. He notes that later medieval philosophy developed new views of causality and substance in the wake of the philosophy of William Ockham and John Buridan, which laid the foundation for the development of the mechanical philosophy and new physics of nature that came to define modern science in the 17th century. He argues that these views themselves could have been influenced by an original theory of matter developed by the Arabic philosopher Averroes in his *Physics* that could be characterized as 'Aristotelian atomism' – a position that could be seen as a bridge between Aristotle's philosophy and the mechanical philosophy. In particular, Lagerlund argues that Averroes' re-reading of an ancient debate over *Aristotle's Physics VII.1* changed the view of the nature of a physical body, and could have played a crucial role in the development of the new conception of corporeal substance which emerged in the European medieval tradition, and paved the way for the mechanical philosophy. Lagerlund's argument suggests that the mechanical philosophy should be seen as emerging out of the Aristotelian heritage, possibly mediated by its culminating development within Arabic philosophy by Averroes, and taken further by thinkers of the Scholastic tradition in Europe.

The next chapter develops a parallel theme of cross-cultural fusion that enriched modern science. In it scientist and historian of Indian science Roddam Narasimha argues that the mathematical approach to understanding nature, a distinctively unique orientation that defined the modern scientific revolution originating in Europe, actually involved a process which combined two traditions from the Bright Dark Ages. The first is the Hellenistic heritage involving the systematic use of deduction based on self-evident axiomatic principles as an instrument for generating knowledge inspired by geometry. The second is the algebraic approach of deploying computational algorithms adopted in Indian mathematics. The Indian advance built upon the computational revolution rendered possible by the Indian decimal place-value number system with zero that led to the development over centuries of powerful algorithms, sophisticated algebraic techniques, equational systems, and even the notion of infinitesimals that paved the way for the calculus. It was combined with the epistemology of computational positivism. Francis Bacon's emphasis on the value of observation, and his notion of *inferred* axioms as first principles of a science, was quite distinct from the Euclidist geometrical orientation of the Greeks that required science to begin with self-evident first principles, and was a creative fusion of Indic and Greek epistemologies. Narasimha argues that we cannot ignore the significant Indian influence on the algebraization of physics and mathematics in the works of Descartes, Newton and Euler, and the ways in which this Indian algebraic orientation came to be fused with the ancient Greek geometrical orientation in early modern mathematics – a combination that shaped the subsequent development of mathematics into the twentieth century.

As with Narasimha, the historian Hyunhee Park explores a similar process of historical connections in the transfer of geographic knowledge from Asia to Europe in the late medieval period. She shows how European geographers received new information to revise their geographic conceptions of the world through a number of political channels that gradually connected Europe to the eastern world in the thirteenth through the fifteenth centuries, such as the Crusades and the Mongol empire. Evidence for these exchanges, according to

Park, can be clearly seen in new types of maps that appeared in late medieval Europe and marked breakthroughs, compared to earlier traditional "T-in-the-O" maps, in depicting the wider world including Asia. Her chapter especially focuses on two new types of European maps of the period. An example of the first type is the world map submitted by a Venetian politician to the Pope which renders the Afro-Eurasian continents with more accurate contours than previous maps, and shows the Indian Ocean reaching the Far East. She notes that maps of this type demonstrate possible access to the Islamic cartographic knowledge of the Indian Ocean and Asia. These maps were soon followed by a second more sophisticated type of maps, such as Fra Mauro's map, that adopted the new coastlines, and also added more detailed geographic information of the wider world including Asia, largely based on information from Marco Polo's account. Her chapter concludes that the connections between Europe and Asia in the late medieval period made such cartographic innovations possible, and paved the way for the unprecedented global connections that later came to be forged through European voyages of exploration.

Hans Pols' chapter on the indigenous medical arts of Indonesia, generally known as *jamu*, examines its philosophy and transmission into Europe in the modern era. He notes that *jamu* originated as medical folklore and was generally practiced by women, who transmitted their insights to future generations orally. As a consequence, only very limited material is available to ascertain the nature of these domestic healing practices. Images of the preparation of *jamu* are present on the mural sculptures of the Borobudur. The various sultanates employed individuals renowned for their medical skills; these at times left written evidence of their insights. In the 16th and 17th centuries, when European traders arrived in the Indies, the physicians who accompanied them eagerly collected herbs, spices, and local medical lore, which they appreciated highly and documented extensively. According to Pols the writings of Garcia da Orta, Bontius, van Rheede tot Drakenstein, and Rumphius (nicknamed the Pliny of the East) were read widely and inspired a medical Renaissance in Europe. European physicians in the 19th century were eager to explore Indonesian herbal medicine to enhance their pharmacopeia, so that their investigations have left ample information on the plants used in *jamu* and their modes of preparation. Somewhat later, anthropologists concentrated on exploring the cosmologies associated with *jamu*. Pols concludes that from these various sources of evidence, an impression can be gained of the basic philosophy of *jamu* and the practices that supported it.

In the following chapter George Gheverghese Joseph examines the evidence for the spread and use of Indian numerals in Southeast Asia, and the extent to which Siamese astronomical tables were influenced by Indian practices. He notes that three main seventh century inscriptions have been found as far apart as Sumatra, Cambodia and Vietnam bearing dates written in Indian numerals and using the Indian *Saka* era which began in 78 CE. The use of the cipher and a place value decimal system is confirmed by these inscriptions. This chapter also examines evidence from a Siamese manuscript containing tables and rules for calculating the positions of the sun and the moon. Apart from some remarkably accurate calculations with respect to the motions of the moon, the meridian of the tables was not a place in Siam but 18° 15' west of Siam, near to the meridian of Varanasi in India. Joseph explores the historical connections implied by such evidence relating to the extent to which cultural influences from India spread widely into Southeast Asia. He notes that Indian influence on Southeast Asia is generally seen to end with the arrival of Europeans in the region and the Europeans are seen as transmitting their own ideas to Southeast Asia which either over-layered or displaced earlier Indian influences. But Joseph concludes that such a historical conception fails to acknowledge that some of the so-called modern ideas that entered Southeast Asia from Europe were really Indian ideas assimilated by European thinkers after contact with India.

Arun Bala considers two recent attempts by Floris Cohen and Toby Huff to answer the Needham Question. Both assume that it is important to ask why modern science did not emerge in China or the Islamic civilization since these two cultures had the most advanced scientific traditions before the modern era. They also assume, Cohen explicitly and Huff tacitly, that it is not important to ask why it did not emerge in the Roman Empire or Southeast Asia because neither of these cultures had any advanced science that could have developed into modern science. However, Bala argues that if we view modern science through the perspective of connected histories, so that it is seen as emerging in Europe only because the latter was able to draw on the scientific achievements of Arabic and Chinese civilizations, then we need not presume that the Needham Question is relevant only when we consider civilizations with the most advanced science. Indeed Europe itself is a descendent of the Roman Empire, so that it becomes proper to examine why modern science was born in Europe but not in Southeast Asia. This is the question his chapter attempts to answer. Bala also maintains that his answer shows that, once we take connective histories of science seriously, we cannot assume that it is the comparatively more advanced scientific civilizations that are most likely to create the next major advance in science – the achievement of Europe in pioneering modern science, even though it was not the most advanced prior to the modern era, proves this point.

Unlike Bala who takes the Needham question seriously, the philosopher Mohd. Hazim Shah does not. He argues that applying it to Islamic civilization in the form 'Why Didn't the Scientific Revolution Happen in Islam?' carries the underlying assumption that undergoing the Scientific Revolution is an unmixed blessing of which all other non-Western civilizations have been deprived. Repudiating such a point of view his chapter explores the alternative idea that the Scientific Revolution was not a pure blessing, and that it created certain cultural problems from which Islamic science was shielded. He begins by asking the conventional Needham question as applied to Islamic civilization and reviews various theories that have been put forward, especially by A.I. Sabra, Pervez Hoodbhoy, and Toby Huff. He proposes the tentative hypothesis that the failure of the revolution to take place is not because of cognitive factors (especially since Islamic science had flourished for several centuries), but because of the differences in the relationship between natural philosophy, mathematics and the experimental method in Islam and the West. In this connection Hazim Shah adopts the thesis suggested by Sabra that science in Islamic Civilization is instrumentalist and not realist in nature. He concludes that this holds certain cultural advantages, although it did not allow for the full exploitation of the potential found in the 'rationalized and disenchanted world-view' of modern science, which was the route taken by the West. He suggests that the lessons obtained for understanding these issues are relevant in resolving some of the contemporary problems relating to science, culture and modernity.

In the next chapter the philosopher and sinologist Franklin Perkins takes a "comparativist" approach to Needham's question. He begins with Kant's argument in the *Critique of Judgment* that science requires the assumption of teleology, which relies on there being a unitary and deliberate cause of the world (i.e., God). He argues that while it is obviously false to say that science requires belief in design, modern science may not have gotten off the ground without that assumption. More precisely, he argues that science emerged in a time when direct appeals to God were given up (marked by the rejection of appeal to final causes), but the confidence that the human mind was made to match the intelligible order of the world remained (the legacy of human beings as made in the image of God). He claims that this period marks European "modernity", which was the result of the unusual length of time over which European culture was dominated by theological explanations. Perkins further argues that such a position never emerged in China because Chinese thinkers had a very short early period dominated by theological explanations. Although it gave them a head start in the beginning, it did not allow them to develop long-term confidence that human beings could have absolute understanding of the order of nature – they went straight from a "pre-modern" to a "post-modern" view. He concludes by pointing out the irony that the progress of science itself has helped overturn the views that may have been required for science to develop in the first place.

The final chapter of this volume by Kapil Raj notes that Needham's Grand Question has until now been one of the major determinants in shaping approaches to the rise of modern science and modernity in the West and its absence in the East. However, he argues that by taking a civilizational approach, the terms of Needham's comparison have pitted putatively long-term characteristics of a part of Asia – China – with supposedly general characteristics of Europe. He attempts to reverse this approach by focusing not on any part of Asia taken in isolation, but on the commerce-driven circulation of people, goods and ideas across terrestrial and maritime trade routes, and delineating how this circulation was itself the locus of scientific and technological innovation until well into the nineteenth century. By carefully examining a Chinese text compiled in 1225 by a senior port official in China, and the translation project of Greek science started by the Abbasids in the eighth century, he shows that this "Asiatic mode" of knowledge and material production drew on resources distributed across the region's various contact zones (ports, inland trade cross-roads and their respective hinterlands). He concludes that the Europeans entering the Indian Ocean in the early sixteenth century were in fact joining an already established, reticular, 'Asiatic' mode of knowledge production. Such sustained interactions between civilisations also became significant vectors of knowledge transmission to emerging European institutions of learning.

What is distinctive of the chapters in the current volume is that they put into question some of the central assumptions that inform Needhamian approaches to the history, philosophy and sociology of science. These may be seen in the following. First Needham assumed an ocean metaphor in which he saw earlier traditions of science (or rather proto-science) rooted in Chinese, but also Indian and Arabic-Islamic, civilizations as rivers of knowledge that flowed into early modern science. This historical thesis was complemented by further philosophical and socio-cultural assumptions he deployed to explain why early modern science developed in Europe. Such philosophical influences included the tradition of Greek philosophy, with the special influence of Platonism and its emphasis on the mathematics of geometry, as well as the social context of Christian monotheism that led thinkers to assume that God ordained the laws of the universe.

The essays in this volume force us to rethink Needham's historical, philosophical and socio-cultural assumptions. In the first place the ocean metaphor of knowledge development is directly disputed by Brennan and Lo. They develop the idea of 'evolutionary syncretism' based on biological and language evolution that serves as a philosophical framework for understanding historical processes that precludes the vision of a single ocean into which species, languages, and presumably scientific knowledge flow. From another direction Sigurðsson and Perkins even question the notion of early modern science as an ocean that absorbed rivers of pre-modern science by noting that that the second scientific revolution associated with the quantum theory has many features that correct and transcend early modern science, but nevertheless resonate with classical Chinese natural philosophy. By contrast Shah questions the notion of the need for Islamic science to flow into the ocean of modern science since the latter has been the source of many problems through the disenchanted world it created.

The stress placed by Needham on Greek philosophy and mathematical rationalism to explain the rise of early modern science is also disputed by other contributors to this volume. Sarukkai does this implicitly by showing how and why Indian epistemological ideas can illuminate contemporary philosophy of science. Brown explores whether Chinese organic materialism would have constituted an in principle obstacle to following through Galilean thought experiments which were crucial to the modes of thinking that led to early modern science. He finds no reason to suppose it does. From a different epistemological direction Wee shows the importance of the experimental methodology articulated by the Arabic scientist ibn al-Haytham as a significant influence on early modern science. The Arabic connection is also pursued by Lagerlund who considers it possible that Averroes views on causality and substance could have contributed to the rise of early modern mechanical philosophy. Questioning Needham's emphasis on the unique role of the philosophy of Greek mathematics as under-girding modern science, Narasimha argues that what was distinctive of early modern mathematics was the combination of the Greek heritage of axiomatic principles embodied in Euclidean geometry with the algebraic approach of computational algorithms developed in India. Indeed taken together these essays reveal that philosophical ideas from many cultures outside Europe not only shaped early modern science, but also illuminate the quantum science today that has gone beyond its original mechanical vision.

The above considerations suggest that we need to go beyond the socio-cultural contexts that informed the Needhamian approach. Needham assumed that civilizations can be treated as tunnels in which knowledge is generated and transmitted to other cultures. Hence, it becomes possible to ask what socio-cultural factors in China, say, obstructed the emergence of modern science, and what other factors in Europe facilitated it. However, many of the contributors here question this central assumption by showing how knowledge exchanges circulated continuously over the Eurasian region millennia before the modern era – a circulation constitutive of the growth of scientific knowledge itself. The essay by Raj adopts a materialist, exchange-based, contextual and contingent approach and draws on recent science studies approaches which focus on the actors, practice, and materiality of knowledge formation and circulation, and concludes that this circulation explains the growth of scientific knowledge.[12] Raj's historical sociology also leads him to question the rivers-and-ocean metaphor that shapes the Needham perspective. Also pursuing the Eurasian connections approach Park argues that it was links with the Islamic and Chinese worlds that led to the vast expansion of geographical knowledge in Europe prior to modern times. Nevertheless, studies that see Eurasian connections as relevant to understanding the emergence of early modern science have generally overlooked Southeast Asia since it is not seen as having achieved a highly developed civilization. The essays by Bala, Pols and Joseph rectify this neglect. Bala asks why, if both Northwest Europe and Southeast Asia were less developed regions on the margins of the Eurasian trading network in premodern times, did circulation from China, India and the Islamic world lead to the birth of early modern science in Europe but not Southeast Asia. Hans Pols also highlights Southeast Asia by exploring the translations and transmissions of *jamu* medical lore from Indonesia to Europe in modern times. Similarly Joseph examines the spread of Indian numerals in Southeast Asia in pre-modern times, so that what are often taken as mathematical imports from Europe into the region are really new imports of Indian numbers after their assimilation in Europe. Such circulations and re-circulations require us to re-examine the notion of civilizations in themselves as incubators of scientific knowledge.

Clearly the authors in this study raise profound questions about the extent to which modern science is directly rooted in the tradition of ancient Greek science as generally supposed, rather than in the science of the Bright Dark Ages. However, what these essays taken together suggest is that the Bright Dark Ages was a period when the scientific traditions forged in many Eurasian civilizations came to intimately interact with each other. Indeed, many of the papers in this volume show that it was the circulation of knowledge across these different early civilizations – sometimes labeled as Axial Age civilizations

---

12   See Latour (1988) and Shapin (1996).

(Eisenstadt 1986) – which produced a new creative age for science not only in the Bright Dark Ages, but also in the modern era. Hence, it is hoped that this volume would pioneer further studies of how dialogical encounters in the Bright Dark Ages across different civilizations not only shaped the growth of science in Islamic-Arabic, Chinese and Indian civilizations, but also in the West.

Bibliography

Al-Andalusi, Said (1991). *Science in the Medieval World: "Book of the Categories of Nations."* Trans. Semaan I. Salem and Alok Kumar. Austin: University of Texas Press.

Al-Biruni (2005). *India: An Account of the Religion, Philosophy, Literature, Geography, Chronology, Astronomy, Customs, Laws and Astrology about AD 1030.* Vols 1 and 2. Trans. Edward C. Sachau. New Delhi: Munshiram Manoharlal Publishers Pvt. Ltd.

Bala, Arun (2006). *The Dialogue of Civilizations in the Birth of Modern Science.* New York: Palgrave Macmillan.

Bala, Arun & George Gheverghese Joseph. "Indigenous Knowledge and Western Science: The Possibility of Dialogue," *Race and Class* (2007) 49: 39–61.

Bayly, C. (2003). *The Birth of the Modern World: Global Connections and Comparisons, 1780–1914.* Oxford: Blackwell.

Cohen, Floris (2011). *How Modern Science Came into the World: Four Civilizations, One 17th-Century Breakthrough.* Amsterdam University Press.

Deming, David (2012). *Science and Technology in World History: The Black Death, the Renaissance, the Reformation and the Scientific Revolution.* McFarland.

Distin, Kate (2010). *Cultural Evolution.* Cambridge: Cambridge University Press.

Duara, Prasenjit, Viren Murthy and Andrew Sartori, (eds.) (2014). *A Companion to Global Historical Thought.* John Wiley & Sons Ltd.

Duhem, Pierre (1985). *Medieval Cosmology: Theories of Infinity, Place, Time, Void and the Plurality of World.* Trans. R. Ariew. Chicago: University of Chicago Press.

Eisenstadt, Shmuel (ed.) (1986). *The Origins and Diversity of Axial Age Civilizations.* Albany: State University of New York Press.

Elman, Benjamin A. (2009). *A Cultural History of Modern Science in China (New Histories of Science, Technology and Medicine).* Harvard University Press.

Goody, Jack (2006). *The Theft of History.* New York: Cambridge University Press.

Garrett Hellenthal, George B.J. Busby, Gavin Band, James F. Wilson, Cristian Capelli, Daniel Falush, Simon Myers. "A Genetic Atlas of Human Admixture History," *Science* (2014) vol. 343, no. 6172: 747–751.

Hannam, James (2011). *The Genesis of Science: How the Christian Middle Ages Launched the Scientific Revolution.* Regnery Publishing.

Hobson, John M. (2004). *The Eastern Origins of Western Civilisation*. Cambridge University Press.

Huff, Toby E. (2003). *The Rise of Early Modern Science: Islam, China, and the West*. Cambridge: Cambridge University Press.

Joseph, George Gheverghese (2011). *The Crest of the Peacock: Non-European Roots of Mathematics*. Princeton: Princeton University Press.

Kuhn, Thomas (1970). *The Structure of Scientific Revolutions*. Chicago: University of Chicago Press.

Latour, Bruno (1988). *Science in Action: How to Follow Scientists and Engineers*. Harvard University Press.

McClellan, James E. & Harold Dorn (2006). *Science and Technology in World History*. Johns Hopkins University Press.

McEvilley, Thomas (2001). *The Shape of Ancient Thought: Comparative Studies in Greek and Indian Philosophies*. Allworth Press.

Mesoudi, Alex (2011). *Cultural Evolution: How Darwinian Theory Can Explain Human Culture and Synthesize the Social Sciences*. University Of Chicago Press.

Needham, Joseph with Wang Ling (1959). *Science and Civilization in China*, Vol. 3. Cambridge University Press.

Needham, Joseph (1970). *Clerks and Craftsmen in China and the West*. Cambridge: Cambridge University Press.

———— (1979). *Within the Four Seas: The Dialogue of East and West*. Toronto: University of Toronto Press.

———— (2004). *Science and Civilization in China*, vol. 7(ii), *General Conclusions and Reflections*. Cambridge: Cambridge University Press.

O'Brien, Patrick. "Historical Foundations for a Global Perspective on the Emergence of a Western European Regime for the Discovery, Development, and Diffusion of Useful and Reliable Knowledge," *Journal of Global History* (2013) 8 (1): 1–24.

Ohashi, Yukio. "Astronomy: Indian Astronomy in China," *Encyclopaedia of the History of Science, Technology, and Medicine in Non-Western Cultures (2nd edition)* edited by Helaine Selin. *Springer* (2008), pp. 321–324.

Pingree, David (1992). "Hellenophilia versus the History of Science," *Isis* 83(4): 554–563.

Pomeranz, Kenneth (2000). *The Great Divergence: China, Europe, and the Making of the Modern World Economy*. Princeton: Princeton University Press.

Raj, Kapil (2007). *Relocating Modern Science: Circulation and the Construction of Scientific Knowledge in Southasia and Europe*. New York: Palgrave Macmillan.

Renn, Jürgen (ed.) (2012). *The Globalization of Knowledge in History*. Berlin: Max Planck Institute.

Ronan, Colin (1983). *The Cambridge Illustrated History of the World's Science*. Cambridge: Cambridge University Press.

Rufus, W.C. (1939). "The Influence of Islamic Authority in Europe and the Far East," *Popular Astronomy* 47(5): 233–238.

Sarma, Nataraja. "Diffusion of Astronomy in the Ancient World". *Endeavour* (2000) 24: 157–164.

Shapin, Steven (1996). *The Scientific Revolution*. Chicago: University of Chicago Press.

Steinmetz, George. "Comparative History and its Critics: A Genealogy and a Solution," in Duara et al. (2014), pp. 412–436.

Varahamihira (2011). *The Brihat-Samhita of Varaha Mihira*, volumes 1–2, translated by N.C. Iyer. Ulan Press.

CHAPTER 1

The Descent of Theory*

Andrew Brennan and Norva Y.S. Lo

1   Introduction

The history of science is beset by difficulties of principle and by problems due to the complex nature of what is studied. On the one hand are questions about the very methods to be used in mapping the growth and development of the sciences, in the face of serious disagreement over what constitutes "science", and how its construction relates to other forms of knowledge, particularly the impressive forms of knowledge found in a variety of cultural traditions, whether Chinese, Islamic, Hindu, Greek, Babylonian or Egyptian. Within each of these traditions there is subtlety and variety: the major cultures are not like uniformly flowing masses of water, but are instead like broad rivers with their various streams, eddies, tributaries, diversions and back currents (Brennan 2004). Attempts at translation and interpretation across cultural boundaries are also beset with difficulties about commensurability of core concepts and ideas. In this context, studies of the growth, development and interplay among various scientific traditions are fraught with hazard.

Some recent constructions of science have been critiqued for their "Whiggishness" – borrowing a term from the political historian Butterfield (1965). From the Whig point of view, the history of science has been a progression over time towards a sophisticated contemporary science (just as, for the Whig historian, the political evolution of various societies can be supposed to have tended towards the enlightened liberal state as ultimate, and preferred, destination) (Mayr 1990). Such Whiggishness is often combined with a form of nativism (another political concept) whereby local traditions and indigenous ways of knowing are lauded for their anticipation of later developments found in the mature phase of a science. Sometimes this involves mainly a recital of supposedly prior achievements and anticipations of later discoveries in unexpected places, such as the ancient Chinese medical uses of ephedrine (and pseudoephedrine), Ibn Qurra's ninth-century work on integral sums and his

* The authors are grateful to Andrea Marcelli whose thorough review and commentary on this paper led to significant changes throughout the text.

anticipation of Fermat, and the pioneering contributions to physics and optics of Alhazen (see Po and Lisowski, 1993, Roshdi 2001, 202–3, and Shuriye 2011). In even more extreme forms, the combination of Whiggishness and nativism may make it seem desirable to search for a 'decoding' of ancient works, such as the *Rig Veda* so as supposedly to extract surprisingly contemporary physics and mathematics from them (see the critique of "Vedic physics" in chapter 4 of Nanda 2003). Whiggish and nativist histories have been put forward as anti-imperialist, and as a rejection of the colonising discourse of mainline histories of science. As such, they may be taken up as contributions to political and social movements, in an intellectual climate that is suspicious of absolute truth, rationality and dominant colonising and imperialist discourses.

At the same time, questions of historical blending or mixing, borrowing and modification, are not well addressed in discussions for and against Whiggish and nativist approaches to the history of science. There is an interesting debate in anthropology over the use of syncretist accounts of change (particularly in the anthropology of religion, see Glazier 2006). Despite the attempt of institutions and devotees to keep religious and other cultural practices as "pure" as possible, historians are able to track systematic borrowings and transformations in these practices over time. For anthropologists, it proved difficult to find a suitable vocabulary in which to described such processes. Terms such as "creole", "hybrid" and even "syncretic" itself need to used warily given their racist and colonial overtones (see Stewart 1999 for a detailed discussion of this matter and a defence of the term "syncretism"). For us, the key features of syncretism is the notion of merging, so that syncretic histories of science and culture will expect different sets of beliefs and practices to merge, cross-fertilize (or cross-contaminate) and in so doing produce new sets of beliefs as part of the general pattern of contiguity and isolation of groups (whether in the geographical or cultural space). As we will see later, such syncretism fits well with an evolutionary view of theory. For the historian looking for evidence of syncretism, the nativist may sometimes provide a useful resource, by suggesting a plausible source for an approach, a theory or a finding that lies outside a particular cultural or theoretical frame.

In this study, we explore the use of evolutionary analogies that might be useful in tackling debates about the circulation and transmission of knowledge, and making sense of the development of the set of practices, findings and theories that are meant to be collected under the label "science". While doing so, we try to avoid falling into either Whiggish or nativist modes of conceptualising the history of science which, in our view, is not always a history of revolutions, but instead can properly be seen, in some cases, to involve evolution and a variety of borrowings from earlier traditions. One challenge is to see how far the evolutionary metaphor can extend to cases where there are not uniform units of selection. If changes in theoretical lineages are regarded as evolutionary to some extent, then does the analogy with biological or linguistic evolution provoke questions to which a prudent history of science is able to give answers? We suggest a tentative and affirmative answer to this question.

2   Flowing Rivers

In an often-quoted remark, Joseph Needham described the confluence of Eastern and Western medieval thought by writing that "one can well consider the older streams of science in the different civilisations like rivers flowing into the ocean of modern science" (Needham 2004: 25; 1970: 397). In keeping with this Whiggish metaphor, Needham distinguishes two different areas of contact. First is the fusion point, where two streams blend, mix and become one. Second is the transcurrent point where one tradition clearly becomes more successful than another in terms of predictions, explanations and fruitful hypotheses. For Needham, the fusion point of Western and Chinese mathematics, astronomy and physics was the early seventeenth century, botany in the late nineteenth, while for medicine no fusion had occurred even to the time he was writing (Needham 2004: 28–31). Indeed, as Needham pointed out, there is a peculiar difficulty in translating terms of art from Chinese medicine, a difficulty that may in some ways parallel the earlier efforts of those who attempted to translate Greek, Persian, Syriac and other medical texts into Arabic. Recent work on the latter case has argued that the attempt at "close translation" should be abandoned in favour of an attempt to capture the meaning of the original – the basis for the "semantic" translations of Abu Zayd Hunayn bin Ishaq al Ibadi (808–873) (also known as Joannitius Onan) (Gorini, 2006). But how much sense is to be made of this form of translation as anything different from appropriation and possibly (mis)interpretation?

If fusion is a borrowing or absorption from one lineage into another, then it is likely to be accompanied by change, modification and transformation, whether the ideas involved are conventions in art, details of religious traditions, or ideas drawn from science and medicine. For a study of a striking case drawn from religion, see Behl's work on whether Mūbad Shāh performed "nothing more than the office of a mere translator" in the Dabistān (Behl 2010: 215). As Behl argues, there are both "clandestine and open agendas" in comparative works and we are well advised to consider the background and even unconscious agendas of the author of any comparative work. Here, then, is a preliminary warning, cautioning against expecting too much from theories of blending and fusion, or analogies with rivers and currents. While rivers and currents are all made of the same stuff, the potential incommensurability of key concepts in different traditions raises a serious question not only over whether blending and fusion are possible, but also over what is meant by the very idea of "blending" itself.

Unhindered by such qualms, Needham claimed that times of fusion were ones of borrowing, adapting and blending the regional discoveries with the ecumenical or universal science. The latter is supposedly one to which all regions and practitioners have access and whose principles of observation, hypothesis-testing and confirmation are open to scrutiny and test by anyone anywhere. Such fusion, Needham observed, will often take place after the transcurrent point has passed. While Western mathematics, astronomy and physics had become superior to Chinese versions by the early seventeenth century, it was only by the middle of the century that the important work of Chinese astronomers had been fully absorbed into the European canon, and in the case of medicine, while the transcurrent point had been reached by the beginning of the twentieth century, there is still no clear evidence of fusion. For example debate still rages to this day on whether acupuncture has anything other than a psychosomatic effect, whether "fake" acupuncture is as good as the "real" thing, and whether electro-stimulation is better or worse than manual acupuncture (Mayor 2013: 414). Needham's diagram detailing some of these fusion and transcurrent points is reproduced below for later reference.

Leaving religion and art to one side, and restricting ourselves only to science and technology, the failure of commensurability of key terms in each tradition's work is likely to be important in medicine, botany and other areas where traditional schemes of classification diverge dramatically from the standard taxonomies of contemporary science.[1] Reflecting on this indicates a problem with Needham's metaphor of many rivers pouring into a single ecumenical sea of science. This is not an issue just for questions of East meets West, but also in understanding local and regional traditions and ways of knowing, even when we confine attention to these within a single social and cultural tradition. What if there are several seas, lakes, and marshes, and what if some of the rivers are best imagined not in terms of fusion, but in terms of sharing some of their courses before separating and feeding different lakes or seas?

In the remainder of the present paper, written from a philosophical rather than a historical perspective, we suggest a change of metaphors and then –

---

1   Some have argued that indeed there is no way of crossing the epistemological break between proto-sciences and their later mature forms, see Canguilhem (1988).

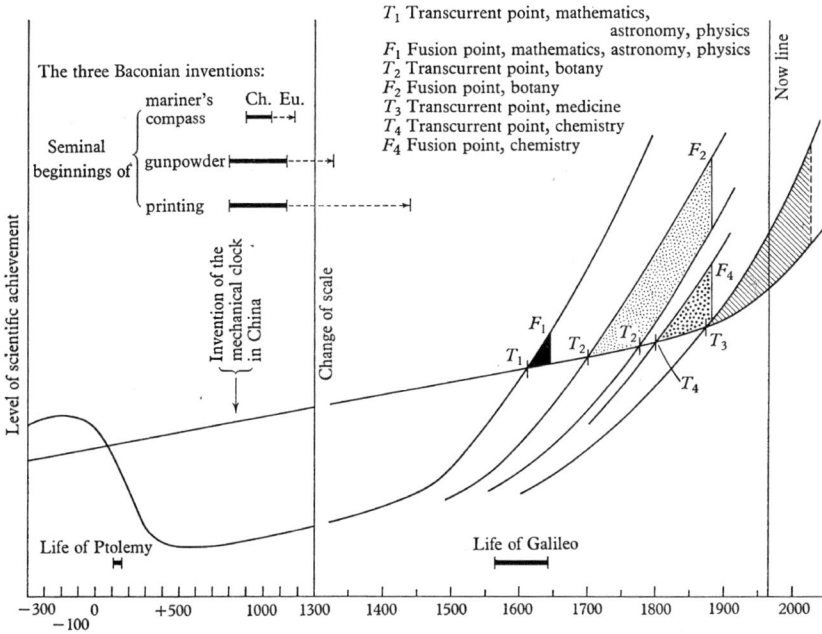

T_1 Transcurrent point, mathematics,
                astronomy, physics
F_1 Fusion point, mathematics, astronomy, physics
T_2 Transcurrent point, botany
F_2 Fusion point, botany
T_3 Transcurrent point, medicine
T_4 Transcurrent point, chemistry
F_4 Fusion point, chemistry

Now line

The three Baconian inventions:

mariner's   Ch. Eu.
compass

Seminal
beginnings of gunpowder

printing

Level of scientific achievement

Invention of the
mechanical clock
in China

Change of scale

Life of Ptolemy

Life of Galileo

−300   0   +500   1000 1300   1400   1500   1600   1700   1800   1900   2000
−100

FIGURE 1.1   *Schematic diagram showing the roles of Europe and China in the development of oecumenical science (Figure 99 from Needham 1970: 414).*

even more speculatively – cover a number of suggestions arising from that change. There is the problem of commensurability for Needham's approach, and – as already noted – the metaphor of rivers and seas is not apt for discussion of cross-fertilisation and the modifications that are inevitable in cases of translating and interpreting. Such a metaphor also hides the cultural, social and religious influences on the politics of syncretism (Stewart 1999, Glazier 2006). For an informed discussion of types of syncretism in the religious context see Budin 2004. It is noteworthy that Budin explicitly uses evolutionary metaphors throughout her treatment of how the goddess Aphrodite 'emerges' from other goddesses around whom there presumably were cults in the ancient world, as when she writes:

> The common belief is that Aphrodite emerged out of Ashtart, thus Ishtar → Ashtart → Aphrodite. In this way, Ashtart and Aphrodite were always closely linked, as the one is virtually the daughter of the other. In contrast to this, I argue that while Ashtart and Aphrodite did evolve from Ishtar, this was a parallel development...; both goddesses evolved

separately, developing their own distinct personae along the way. (Budin 2004: 110–111)

Even if evolutionary metaphors are themselves ultimately inadequate, their use in studies of syncretism and other comparative work suggests they have the power to take us some way beyond Needham's approach, and enable us to appreciate new ways of thinking about the Needham question of why the modern sciences emerged when and where they did. We will look at analogies with both biological and linguistic evolution, taking these as the first step away from the rivers, seas and lakes metaphor. So we start by investigating the extent to which the modern sciences can be thought of as having a place on an evolutionary tree.

3   The Struggle for Existence

If evolution is defined as the change in the distribution of inherited characteristics in a population over time (Endler, 1986), then there are normally said to be three key ingredients. These are variation, multiplication and heredity (Hodge 2009: 243 and O'Brien et al. 2003). In any species population, each member varies slightly from the others, a feature that – some time after Darwin – was attributed to small genetic variations across individuals. Such variations are often of little or no significance, though sometimes they affect the longevity or reproductive success of some individuals compared to others in the same species. The key idea in plant and animal evolution is that individuals strive to multiply by reproducing. Darwin himself makes the point that so rich is nature's capacity in this regard that many more individuals are produced than can possibly survive. The result is a struggle for existence, one that is at its most intense within the species, and less intense among individuals of different species, since the latter may well depend on rather different resources for their survival and reproduction. Hence we find constant diversification within individual species:

> *Therefore* during the modification of the descendants of any one species, and during the incessant struggle of *all species to increase in numbers, the more diversified these descendants become, the better will be their chance of succeeding in the battle of life.* Thus the small differences distinguishing varieties of the same species, will steadily tend to increase till they come to equal the greater differences between species of the same genus, or even of distinct genera. (Darwin 1859: 127–128)

Evolution – the changing distribution of heritable traits within populations – ultimately brings forth speciation, the emergence of a diversity of species. Darwin's original, gradualist, account of speciation required the isolation of plant and animal populations, the appearance of slight variations from one individual to another and the passing on of characteristics through heredity. In time, through the accumulation of slight changes favoured by specific environments (adaptation), isolated species populations would develop in slightly different ways from each other. Starting with one original species, this process would in the end lead to the formation of new varieties, new sub-species, and – after enormous periods of time – completely new species. For punctuationists who deny that evolution is always gradual, the Darwinian account just given would, of course, simply be one kind of evolution among others (see Gould and Eldridge 1977). The concept of evolution, as such, does not rule out the existence of periods of rapid change.

In a remarkable parallel to the theory of biological evolution, nineteenth century linguists aimed to plot similar processes of evolution in languages. August Schleicher (1821–1868), for example, argued that it was possible to draw up a family tree – a *Stammbaum* – for many of the languages of Europe and Northern India which would show they were descended from a single ancestral language (called 'Indo-European'). While recent critics of Schleicher have criticised the "romantic" and Hegelian turn that he gave to the notion of evolution (see McMahon and McMahon 2013: 7), there is a striking analogy here. Just as species form families, while within species there are sub-species and varieties, so languages form families, and within languages there are dialects and regional variants. The brothers Grimm, well known for collecting folklore and fairytales, were among the group of German philologists who worked out principles describing the history of consonant-shifts over time, shifts that show how Latin *quod* became English *what*, or the Sanskrit *bhrātṛ* became the Norwegian *broder*, German *Bruder* and English *brother*. While biologists studied the fossil record, comparative anatomy and embryology to help in plotting the descent of species from earlier forebears, linguists studied ancient texts, the dialects of isolated rural communities and patterns of consistent sound change to explain the development of different languages from common ancestral forms.

Linguistic evolution, however, does not quite match the biological case. Some would distinguish historical from genetic evolution as the explanation of this:

[W]e must constantly remind ourselves that we are dealing here with evolution as a metaphor: variation and change can be interpreted as

analogous to genetic mutation, variation and selection, but in our terms
the rise of linguistic variants and their possible embedding in language
systems through change are crucially historical rather than evolutionary
processes, which therefore do not in fact involve any genetic mutation,
variation or selection. (MacMahon and MacMahon 2013: 14)

Such a stance takes literalism too far, in our view, and undervalues the role of
metaphor in the history of science. While modern genetic theory has added
depth to Darwin's original account of evolution, it is clear that even in the
absence of an account of what the units of selection were supposed to be,
Darwin had developed a significant and startling explanatory theory. This is
not to deny that there are problems in applying the notion of evolution to languages. While different species are commonly unable to breed successfully with
one another, languages show immense capacity for cross-borrowing. Even an
analytic language, like Cantonese, shows borrowings from English, including
appropriations of idioms, that reflects the long period of British colonialism
during the twentieth-century (personal observations by the authors). Given
the possibilities of cross-language fertility, it might seem that, if the idea of
languages as species is taken seriously, then most or all of the world's languages
are in effect subspecies, that is members of one huge family, rather than collections of truly separate lineages (see Croft 2006: 105–106). This is a disputed
matter in theoretical language studies. None the less many researchers find
it plausible to maintain that if a species is a population whose members can
interbreed, then a speech community is a group whose members can interact
linguistically with one another and are to a degree communicatively isolated
from other such communities (see Dixon 1980, Chambers and Trudgill 1998,
Croft 2006).

   Where do scientific theories fit in these models? Are they more like biological species, unable to fuse with each other in fertile ways? Or are they
more like natural languages, capable of productive hybridisation, creolisation
and cross-breeding – if we dare to use the very terms over which we earlier
sounded a warning? Using the biological analogy, some of the Darwinian metaphor appears apt. As long as theories and ideas are passed from one person
to another, and not changed too much in the process, there is heritability. But
there is variation even in simple situations, for it is seldom the case that two
people enjoy anything like perfect communication. It is well known that a
story passed through a group of people, one at a time, is very different in its
ninth or tenth telling from how it was at the original telling. It is a posit of
many communication theories that in sending a message, the originator has
to encode it for transmission to the receiver, who then decodes it by means

of the same code in which the message was sent. Problems of encoding and decoding can explain the transformation of the message content within just a few repetitions. Although beyond the scope of the present study, it would be interesting to explore the role of misconception, mistransmission and misunderstanding in the history of scientific creativity and theory change. A similar remark would apply to the study of syncretism in religion where mistakes in transmission may have a role to play in describing changes and dislocations of meaning within and between traditions.

Leaving aside Behl's points about the hidden agendas of translators and interpreters, we now see an immediate problem about the transmission of ideas, concepts, clusters of these, and certainly of whole theories themselves. If languages were perfectly determinate, or built of determinate components such as genes, then there would in theory be few transmission errors, and communication would work much better than we normally experience. But language is an imperfect coding system, so there can be many slips between the transmission and the receipt, even in the case where the parties involved are speakers of the same language. If we assume each person has their own unique language, one that varies ever so slightly from every other speaker's version of the language – what linguists and philosophers call the individual's *idiolect* – then discussions of ideas are bound to be fraught with multiple possibilities of ambiguity and mutation. Even in the setting of a class, a committee meeting, or other formal gathering, it is likely that each participant will leave with a slightly different understanding of what has been discussed, achieved or agreed. There is a biological analogy to all this: each individual in a population is phenotypically distinct from all the others (and even identical twins reveal phenotypic plasticity). Each idiolect within a language is likewise distinct from every other idiolect, though here the analogy between genome and language breaks down. It would be implausible, even if harmless, to regard every speaker of a language as carrying the complete vocabulary and grammar of the language while their actual language only "expresses" some parts of the language.

Despite its problems, the analogy now shows some interesting things. As with intra-species variation, many of these indeterminacies may be of little consequence. But in some cases, two people who think they share an agreed theory or view may well express different varieties of the theory in question. The more complex and elaborate the matter, the greater the likelihood that there is diversity of understanding in a group, even in cases where there would seem to be a clear set of codings using mathematical or logical symbols. When the agents engaged in communication come from different linguistic backgrounds, the chances for diversity rise and the chances of new variations on an idea increase quite dramatically. This is an often overlooked source of scientific novelty: in a dialogue, new ideas can be formed that are not the product of either participant alone. Rather, a good dialogue can be transformative, a partnership in which unexpected novelty emerges in ways not anticipated by the participants. A hermeneutics of science, just as much as a hermeneutics of education, should leave space for just such novelty:

> Words have effective histories we do not control. We remain for the most part unaware of the sub-surface philological currents which both shape our mental shore-lines and always threaten to draw us into new cognitive waters. An unaccustomed juxtaposition of words, a novel locution, an accidental remark can suddenly breech the mental and social dykes of habit which channel the movement of customary usages of meaning. (Davey 2011: 50–51)

When we turn to speciation – the emergence of new lineages – the situation becomes more complicated. Needham's picture of the emergence of the one ecumenical science ignored the possibility of diversity just discussed. If each thinker, transmitter and student is already the locus of a slightly different expression of a theory from every other, then each person, is a potential parent of a new theoretical lineage. Consider the parallel case in evolution. Species lineages are normally shown on a tree structure, with branching showing how later lineages evolve from earlier ones. Now think of a new branch growing on such a tree. A bold way of conceptualizing this change is to argue that there is a point mutation at the node joining the new branch and that mutation characterizes the first member of the new species. As one writer has pointed out, "...this assumption is a caricature of biological reality", yet it can be useful when working on theories of population genetics (Kopp 2010: 565).

That is why each individual is a potential founder of a new species. If the caricature is taken seriously, it would mean that there are millions of potentially new species appearing all the time – most of them being very short-lived indeed. The practical uselessness of such a way of thinking about species novelty has led many biodiversity theorists to query such a way of conceptualizing speciation (Rosindell et al. 2010, Kopp 2010). Instead, some theorists have argued that short-lived lineages are best thought of as variations within a longer lineage, with new species emerging only after a suitable number of generations (Kopp 2010: 567). There is a moral here for history of science, one that reminds us of the importance of the environment – of schools, constellations of thinkers, and institutions for preserving and transmitting knowledge. Think of outstanding characters in the Chinese tradition, for example the eleventh-century experimentalist Shen Kuo (沈括). Although a marvellous polymath, whose work was strikingly advanced for the time, Shen Kuo had no school around him, and left few intellectual descendants to follow in his footsteps, apart from his contributions to navigation and astronomy. He was like a point mutation, who could have been the origin of one or more lineages – but it was not to be. Commentators have pointed to the lack of systematicity in his thought, the burning of some of his books after his death, and a number of other factors that could explain this (see for example Huff 2003). But for our analysis, his isolation, and the lack of a constellation of other thinkers around him, are key to explaining his failure in founding a scientific lineage. In this way, environmental factors are just as vital in explaining the birth of scientific lineages as they are in telling us about biological ones. Figures like Shen Kuo are found in many traditions, perhaps showing that faced with similar problems, thinkers at very different places and times will come up with similar approaches to their solution. Moreover, the isolation of such individuals – like their lack of an institutional context within which to have influence and pass on their contribution to subsequent generations – means that there will be many cases where there was potential for establishment of a lineage, but no lineage emerged.

This detour into theories of speciation throws up a note of caution. A working hypothesis for biologists is that common structures are often, but not always, evidence of common descent. Comparative anatomy was at the centre of Darwin's claims, but similar structures in science and technology need not imply a history of contact between two cultures. Likewise, in the case of parallel evolution, we now know that genetically different organisms develop similar responses to similar environmental conditions (see Shelley 1999: 67–68). In the case of languages, there is no agreement among linguists about whether all human languages share a common descent, although there have been claims that certain traits are common to all languages (Whalen and Simons 2012: 155). Instead, there appear to be over 100 language families worldwide (Lewis 2009). Given the observation above about borrowings, it may seem that these families in turn are best thought of as belonging to a super-family, and that borrowings and cross-fertilisations are always possibilities for each member of the superfamily.

Where does the account of the descent of theory sit in relation to these possibilities? One approach would be to look for a single history of Needham's "ecumenical science", or some small number of sciences, a history that can be traced to the oldest common ancestor. Such indeed is the trajectory of many standard histories of Western science. If we restrict our attention to the physical and natural sciences, excluding medical theory and practice, then there is a familiar story of descent from the Egyptian and Babylonian common ancestor. According to this story, a line of descent runs from ancient Egyptian and Babylonian science, through classical Greek and Hellenistic achievements, and then to the Romans (who were wonderful technologists in their own right,

but also useful to the descent of theory in the way that they preserved the legacy of Greek theoretical speculation). As the classical world fell into ruins, the
preservation of the Greek legacy passed to the expanding Islamic empire to
such an extent that for some centuries – the Bright Dark Ages – Arabic was
the language of the sciences. The legacy of Islamic science (using "Islamic" in a
geopolitical, rather than religious sense), then passed back to the West over the
course of the Renaissance and following centuries, where – as Needham's diagram shows – there were a series of revolutions in physics, astronomy, botany,
chemistry and other natural sciences (see Figure 1 above).[2]

An alternative way of conceptualizing what took place is to think of families of theoretical and speculative lineages that form networks in their own
right, but which came into contact with each other through trade, war, political expansion and so on. If the medical sciences and technologies are included
in this picture, then there will be different clusters of lineages marked by the
notions of Hindu science (using "Hindu" as a term of geopolitical, not religious,
identification) and Ayurvedic medicine, Chinese science, technology and
medicine, the Greek medical tradition, the Islamic, and other families, some
of whom have very long lineages. As we go back in history, we come to the
same problem faced by students of syncretism, such as Budin: we have to work
from fragmentary evidence, isolated historical and archaeological finds and
other scraps from which theories of the early stages of scientific evolution are
put together.

The resulting picture neatly captures two themes – connectivist versus
comparativist views of the history of science. The connectivist aims to create
lengthy lineages traceable to earlier ancestors, in turn traceable to still earlier
ones, and so on, but at the risk of theorising beyond the available data. For the
connectivist, the existence of chain pumps in Mesopotamia, Rome, China and
the Islamic world calls for investigation of ways by which this technology could
have derived from a common ancestral technique possibly Chinese (see Mays
2010, Tamburrino 2010). The more wary comparativist may reflect that the
same problem provokes similar solutions in different places, and be reluctant
to conclude, in the absence of archaeological or other historical evidence, that
similar technologies imply contact or circulation of ideas. One worry is that by
merging traditions into causal chains of influence (Babylonians, Greeks, Arabs,
etc.) the connectivist and syncretist threatens the autonomy, creativity and
sparkle that each tradition has in itself. To allay such a fear, a sensitive version of connectivist thought would eschew the idea that one lineage has value
only to the extent that it contributes to another, more recent, one. An equally

---

2   For a classic work on the importance of the Renaissance, see Garin (1941).

sensitive comparativist would argue that the exchanges, trades, wars and other dialogues among lineages spark forms of novelty that cannot be entirely credited to one tradition or another. As Arun Bala argues it may require the meeting between Ghazali's critique of the rationalistic apriorism of Greek science and the organic materialism underlying Chinese engineering for a properly self-critical and empirical modern science to emerge (Bala 2006: 119–30). Viewed in this way, a kind of evolutionary syncretism may be able to provide a synthesis of both the comparativist and the connectivist approaches, one that is sensitive to evolutionary processes, accidental cross-fertilisations and also forms of unintended influence that have more in common with causal rather than cultural processes.

The cultural, political and ideological issues here raise concerns of great sensitivity. To value the dignity and autonomy of diverse traditions seems like an appropriate orientation to cultural diversity in the contemporary world. While the comparativist has no problem meeting this challenge, the connectivist has to argue that there are forms of novelty that would likely not emerge in the absence of a real dialogue between separate grand lineages. The metaphor of evolution provides, we maintain, a helpful way of thinking about these issues, one that is not committed to finding a single common ancestor for modern science, nor that imposes a highly Eurocentric view of the topic.

4   Conclusion: Scientific Theory and Christendom

We conclude the present paper with a provocation, one that returns again to the vexed issue of why modern science emerged when it did and where it did. Let us look back at Figure 1 above. Much of European history can only be understood in terms of the history of one religion, namely Christianity. Indeed, one eminent historian of Europe points out that it was only in the sixteenth century that people began to think of Europe as a geopolitical entity. Before that, the continent was very much defined as the domain of Christianity – Christendom. Only when the established churches split from each other, and new versions of Christianity began to take form did Europe itself emerge into people's thinking:

> Until the 1530s, Christendom had been split into two halves – Orthodox and Catholic. From the 1530s onwards it was split into three: Orthodox, Catholic and Protestant. And the Protestants themselves were split into ever more rival factions. The scandal was so great, and the fragmentation was so widespread, that people stopped talking about Christendom, and began to talk instead about "Europe". (Davies 1997: 494–496)

Correcting Needham's diagram in light of this, we could treat the line labelled "Europe" as a mistake. There should be two lines there: one for Graeco-Roman science (Ptolemy after all was both Greek and Roman), and that line should probably not decline so sharply after the first century. A second line, labelled "Christendom" could then be identified, following the trajectory of the remainder of Needham's line for Europe. This relabeling makes more sense of the ideological situation in Christendom itself, and its relations – including wars – with the caliphates. In making this claim, we are not arguing that only the existence of an ideology is relevant to identification of a scientific community. As the case of chain pumps shows – and that of dietary prescriptions common to the various ancient medical traditions – material capabilities are also highly relevant to the identification and comparison of different scientific communities.

With the diagram changed, it is of course obvious that the decline of Graeco-Roman science matches the decline of the Roman Empire and the end of the classical period. The line for Christendom is now parallel to the line for China up to the early modern period, and it is interesting to speculate how we would graph Islamic science and technology from the seventh to the fifteenth century, a line that is notably absent from Needham's figure. In addition to the well-known theories about agriculture, trade, war and weaponry – all key elements in narratives about transmission, circulation and evolution of knowledge – we would now like to throw in one additional consideration. To do so, we must take issue with one further aspect of Needham's theories, because of the way he combines together the pure and the applied, the science and the technology. The so-called "dark ages" of Christendom were a time when technology in the Christian countries was far behind that in China and the Islamic countries. Yet it would be a mistake to say that theoretical and intellectual life was at a low point. On the contrary, some of the great achievements of Christian philosophical thought were made during that period, from the Church fathers to the time of Aquinas and beyond. And from the medieval to the early modern period – roughly 1000 to 1600 – there occurred an immense flowering of logic. There were also the beginnings of an attempt to find laws of nature on analogy with the supposed laws of magic and ritual.

Medieval thought was inspired by a picture of the human mind that depicted it as similar in important ways to the mind of God himself. By studying intellectual pursuits, and by gathering people together to explore questions of ethics, metaphysics and even logic, Christian thinkers saw themselves as coming closer to understanding God. The mind, or soul, in other words is a divine part of the human, and as Roger Bacon – himself incidentally a student of Alhazen's work – remarks in his early writings the "agent intellect" is directly aware of spiritual beings. In his later work he identifies that form of the intellect with God himself (Hackett 2012). Along with such a view of the worthwhileness of engaging in philosophical and theoretical speculations went an institutional willingness to provide for gatherings of devotees dedicated to the study of theology and also to engage in the study and transmission of logic, grammar, astronomy and other intellectual arts. Certainly, gathering for the purposes of discussion of the affairs of the *polis*, and related matters extending to metaphysics and philosophy was not by any means a Christian invention. Consider, for example, the role of the *agora* in Greek settlements, and the explicit discussion of community and the specialized roles it imposes on its members in Plato's *Republic*. Christianity seems to have inherited a Greek view on the collective pursuit of knowledge, and provided institutions – the monasteries – from which sprang the medieval universities, with their great libraries, collections of manuscripts and commitment to providing education and medical care to the wider community.

So here is an ideological character worth noting: an other-wordly religion that brings people together in worship and study, emphasizes the divinity in humans, and makes pure speculation, including logic and philosophy, one of the highest callings for humans. Those who see Christianity aligned to the rise of capitalism will perhaps recall Tawney's idea that medieval and early modern forms of Christianity were anti-materialist and not in favour of profit for its own sake (Tawney, 1961). In keeping with this anti-materialist stance, the study of theology, philosophy, the principles of mechanics and the movements of the heavenly bodies appear also to have been valued for their own sake, or at least for enlightening the student on the divine purpose.

A related issue concerns the intelligibility of nature. Much has been made of the fact that classical Greek, and also Chinese, thought was correlational, in comparison to the more causally-structured thought of modern Europe. When modern science emerged in the seventeenth century, it replaced correlative cosmological thinking only slowly and with difficulty (see Part IV of Graham, 1989). Now why did the Greeks love magic as much as philosophy, and why did the Chinese with their generally materialist stance regard uncovering some correlations among "the ten thousand things" as being as good as it would get? Part of the answer here may lie in the idea that there are laws and principles that apply likewise to dealings with natural forces and to dealings with gods and demons. But neither the Greeks nor the Chinese had any ideological or theological commitment to the notion that the world had been created by an omnipotent mind of infinite capacity – one that in some respects worked like the rather more limited human mind. For thinkers in Christendom, by contrast, the world was almost guaranteed to be intelligible: for it was the design of a mind whose depths and contours we can at least begin to fathom. The intelligibility of nature, in other words, comes as part and parcel of the view that each human being contains a spark of the divine. If nature is intelligible, then the scientist has an incentive to push on in the quest for discovery, an incentive that those in other traditions can be said to lack. This incentive runs far deeper than simply the commitment to the view that nature is intelligible in the limited sense that some scientific principles are self-evident to a clear and unbiased mind (see Dear 2006: 24, and compare ch. 1 of Craig 2004).

A further point can be made about Christianity in relation to other religious traditions that have had major effects on the culture of the countries in which they hold sway. In political terms religions and their varieties are sometimes described as having either centripetal tendencies – by bringing people together in relatively unified communities – or centrifugal ones – by separating people into opposing groups which can threaten social disharmony and breakdown.[3] In terms of their effects on individuals, religions can be measured on the same scale. The institutions of Christendom encouraged groups of thinkers to get together, and encouraged exchanges among the groups as monks moved from one centre of study to another. Other religions, with an emphasis on individual enlightenment, might be thought to have the opposite, centrifugal tendency. In both Hinduism and Buddhism there is an emphasis on the subject becoming detached from the illusions of the material world and on self-realization as liberation from the world. For the Christian, by contrast, the world is the work of God, as the Psalmist wrote: "The heavens declare the glory of God; and the firmament sheweth his handywork". Uncovering nature's secrets, then, became itself part of the Christian's work, work that can legitimately aim at objective knowledge. The superficial similarity between the different world views can mislead: for example, a sixteenth-century Buddhist and sixteenth-century early modern scientist would both believe that our senses deceive us and that much of what we think in daily life is an illusion. The latter, however, would also think that penetrating the illusion leads us to clearer knowledge of how the world works, and to that extent reveals the mind and works of the creator. Just as biological and linguistic evolution take place within an environment that selects for some developments and not for others, the cultural, social and religious environment of a given period can be expected to play a role in the descent of theory.

For some readers, the line of thought here may recall the Merton thesis about the influence of the Reformation on the development of modern science (see Merton 1938, Cohen 1990, Shapin 1988). We are not defending this

---

3 We owe this idea to a discussion with Christopher Mackie.

view here. Instead, we are suggesting that for the evolutionist this concluding section provides new material for theorising, and yet more reason to be sceptical about the Needham question: *Why did modern science emerge in Europe?* That question makes no more sense than asking why *homo sapiens* emerged in Africa, and admits of no answer better than: *because that is what we now call the place where they emerged.* Would it be better to remove the question's socio-geographic bias, and ask instead: *Why did modern science emerge in Christendom?* Such a change may be a slight improvement on the Needham question, but we need to be cautious in trying formulate an answer. The environment of early modern Christendom was favourable to speculations about an intelligible nature, and also provided a degree of socio-political upheaval that provoked the collision of many traditions and introduced a period of rapid evolution, and cross-fertilisation. Just as contingent environmental factors impinge on biological evolution, social and political factors, including war, religious schisms, colonisation and the Renaissance paved the way for a period of theoretical evolution that was to change forever our understanding of the universe. It was in this way that modern science emerged in Christendom. It did not emerge as a product of Christendom, but instead as a result of the syncretic evolution and timely collision of many traditions from around the globe.

Bibliography

Bala, Arun (2006). *The Dialogue of Civilizations in the Birth of Modern Science*. New York: Palgrave Macmillan.

Behl, Aditya (2010). "Pages from the Book of Religions," in Haberman, D. and Patton, L. (eds.), *Notes from a Mandala: Essays in the History of Indian Religions in Honor of Wendy Doniger*. Newark: University of Delaware Press.

Brennan, Andrew. "The Birth of Modern Science: Culture, Mentalities and Scientific Innovation," *Studies in History and Philosophy of Science* (2004) 35: 199–225.

Budin, S.L. "A Reconsideration of the Aphrodite-Ashtart Syncretism," *Numen* (2004) 51(2): 95–145.

Butterfield, H. (1965). *The Whig Interpretation of History*. New York: W.W. Norton.

Canguilhem, G. (1988). *Ideology and Rationality in the History of the Life Sciences*. Cambridge: MIT Press.

Chambers, J.K. & Peter Trudgill (1998). *Dialectology* (2nd ed.). Cambridge: Cambridge University Press.

Cohen, I.B. (ed.) (1990). *Puritanism and the Rise of Modern Science: Understanding the Merton Thesis*. New Brunswick, N.J.: Rutgers University Press.

Craig, E. (2004). *The Mind of God and the Works of Man*. Oxford: Oxford University Press.

Croft, W. (2006). "The relevance of an evolutionary model to historical linguistics," in O.N. Thomsen (ed.), *Competing Models of Linguistic Change: Evolution and Beyond*. Amsterdam-Philadelphia: John Benjamins, pp. 91–132.

Darwin, C. (1859). *On the Origin of Species by Means of Natural Selection*. London: John Murray.

Davey, N. (2011). "Philosophical hermeneutics: An education for all seasons?" in P. Fairfield (Ed.), *Education, Dialogue and Hermeneutics*, London & New York: Continuum, pp. 39–60.

Davies, Norman (1997). *Europe: A History*. London: Pimlico.

Dear, Peter (2006). *The Intelligibility of Nature*. Chicago: University of Chicago Press.

Dixon, R.M.W. (1980). *The Languages of Australia*. Cambridge: Cambridge University Press.

Endler, J.A. (1986). *Natural Selection in the Wild*. Monographs in Population Biology 21. Princeton, New Jersey: Princeton University Press.

Garin, E. (1941). *Il Rinascimento Italiano*. Milan: ISPI.

Glazier, S.D. (2006). "Syncretism," in H.J. Birx (ed.) *Encyclopedia of Anthropology*. Sage, pp. 2150–2152.

Gorini, R. "The process of origin and growth of the Islamic medicine: the role of the translators. A glimpse on the figure of Hunayn bin Ishaq," *Journal of the International Society for the History of Islamic Medicine*, (2006) 4(8): 44–47.

Gould, S.J. and N. Eldredge. "Punctuated equilibria: The tempo and mode of evolution reconsidered," *Paleobiology* (1977) 3: 115–151.

Graham, A.C. (1989). *Disputers of the Tao*. La Salle: Open Court.

Hackett, Jeremiah. "Roger Bacon" in Edward N. Zalta (ed.), in *The Stanford Encyclopedia of Philosophy (Winter 2012 Edition)*. URL = <http://plato.stanford.edu/archives/win2012/entries/roger-bacon/>.

Ho, P.Y. and F.P. Lisowski (1993). *Concepts of Chinese Science and Traditional Healing Arts: A Historical Review*. Hong Kong: World Scientific Publishing.

Hodge, J. (2009). "Evolution," in *The Cambridge History of Science, Volume 6: Modern Life and Earth Sciences*. Cambridge: Cambridge University Press, pp. 243–264.

Huff, Toby E. (2003). *The Rise of Early Modern Science: Islam, China, and the West*. Cambridge: Cambridge University Press.

Kopp, M. "Speciation and the Neutral Theory of Biodiversity," *Bioessays* (2010) 32: 564–70.

Lewis, M. Paul (ed.) (2009). *Ethnologue: Languages of the World*, Sixteenth edition. Dallas, Tex.: SIL International. Online version: http://www.ethnologue.com/.

MacMahon, A. and R. MacMahon (2013). *Evolutionary Linguistics*. Cambridge: Cambridge University Press.

Mayor, D. "An exploratory review of the electroacupuncture literature: clinical applications and endorphin mechanisms," *Acupuncture in Medicine* (2013) 31(4): 409–415.

Mayr, Ernst. "When Is Historiography Whiggish?" *Journal of the History of Ideas* (1990) 51: 301–309.

Mays, L.W. (ed.). (2010). *Ancient Water Technologies*. Dordrecht: Springer.

Merton, Robert K. "Science, Technology and Society in Seventeenth Century England," *Osiris* (1938) 4: 360–632.

Nanda, Meera (2003). *Prophets Facing Backwards: Science and Hindu Nationalism*. Piscataway, NJ: Rutgers University Press.

Needham, Joseph (2004). "The Roles of Europe and China in the Evolution of Oecumenical Science," originally delivered an 1967, and reprinted in Robinson, K.G. (ed.) *Science and Civilisation in China, vol 7, part 2*. Cambridge: Cambridge University Press (2004). Also reprinted as ch. 19 of Needham (1970).

———— (1970). *Clerks and Craftsmen in China and the West: Lectures and Addresses on the History of Science and Technology*. Cambridge: Cambridge University Press.

O'Brien, M.J., R.L. Lyman, and R.D. Leonard. "What is evolution? A response to Bamforth," *American Antiquity* (2003) 68(3): 573–580.

Roshdi, Rashed (2001). "Chapter 2.1 Mathematics", in Al-Hassan, A.Y., Ahmed, M., and Iskandar, A.Z. (eds.) *Science and Technology in Islam: Part 1 The Exact and Natural Sciences*. Paris: UNESCO.

Rosindell, J., S.J. Cornell, S.P. Hubbell and R.S. Etienne. "Protracted speciation revitalizes the neutral theory of biodiversity," *Ecology Letters* (2010) 13(6): 716–27.

Shapin, Steven. "Understanding the Merton Thesis," *Isis* (1988) 79(4): 594–605.

Shelley, C. "Preadaptation and the Explanation of Human Evolution," *Biology and Philosophy* (1999) 14: 65–82.

Shuriye, A.O. (2011). "Islamic Position on Physics with Reference to Ibn Al-Haytham," *International Journal of Applied Science and Technology*, 1. On-line at http://www.ijastnet.com/journals/Vol._1_No._2;_April_2011/7.pdf.

Stewart. C. "Syncretism and its Synonyms," *Diacritics* (1999) 29(3): 40–62.

Tamburrino, A. (2010). "Water technology in ancient Mesopotamia," in L.W. Mays (ed.), *Ancient Water Technologies*. Springer. pp. 29–52.

Tawney, R.H. (1961). *Religion and the Rise of Capitalism*. Harmondsworth: Penguin Book. Originally published 1926.

Whalen, D.H. & G.F. Simons. "Endangered language families," *Language* (2012) 88(1): 155–173.

CHAPTER 2

Philosophical Implications of Connective Histories of Science

Sundar Sarukkai

1 The Problematic Link between History and Philosophy of Science

The relationship between history and philosophy of science (HS and PS henceforth) has been a difficult one. Both these disciplines, in their development as highly specialized discourses, have had less and less to do with each other. So much so, there have been explicit claims that history of science does not matter to philosophy of science and vice versa. Philosophers often find the extremely detailed descriptions in history of science distracting as it seems to be more about history than about science. Similarly, philosophy of science's concern with finer philosophical details has alienated the attention of a large number of historians of science.

By contrast I will argue that the connective history of science has important consequences for philosophy of science as well as science education. However there are specific points that relate HS and PS (Ariew and Peter 1986, Burian 1977 and Pinnick and Gale 2000). This includes clarification on the definition of science, nature of science, idea of theory and practice, use of science as an ideological (colonial, developmental, national) tool and the impact of this on the legitimization of science, science's relationship with language, particularly mathematics, and so on. We will see how some of these themes get modified when viewed through a connective history of science (CHS henceforth).

The relationship between history and philosophy of science has not been too congenial. Both of them might have the same subject matter but their approaches are sometimes so different that it becomes difficult to draw on one another. This has become a bigger problem as the disciplines become more and more specialized. However, there have always been voices which have tried to, if not integrate, at least put these disciplines on talking terms. For example, Hanson (1962: 580) suggests that 'history of science without philosophy of science is blind' and 'philosophy of science without history of science is empty.' However, in what follows he discusses how these two are not 'logically related' but allows for their mutual relevance.

© KONINKLIJKE BRILL NV, LEIDEN, 2016 | DOI 10.1163/9789004264199_004

What are the possible ways by which HS and PS can be relevant to each other? Nickles (1995: 139) argues that following Kuhn historians of science prioritized history of science by arguing that philosophy of science did not have the resources to solve its problems without drawing upon history of science. Generally, there was an acceptance that historical data add to the weight of philosophical understanding. He points out that history of science has 'greatly enriched philosophy of science by raising new problems and suggesting new solutions' (ibid., 161).

It is generally accepted that Kuhn's history of science has had a profound impact not only on philosophy of science but also on fields such as science education. Hoyningen-Huene (1998) argues that Kuhn's theory 'ran counter to many philosophical convictions about science' at that time and hence was influential to that field. Among the reasons he gives for Kuhn's influence are his critique of scientific progress as well as method, his focus on the historical nature of theories, his critiquing Popper's 'critical rationalism', and his emphasis on the communal nature of science over the individual as the agent of scientific change. Also Kuhn's contribution to the idea of incommensurability as well as to a theory of meaning gave new material for philosophy of science (ibid. 11). In a later paper, Hoyningen-Huene reiterates the argument that the historiography of science cannot be 'entirely philosophically innocent' (Hoyningen-Huene 2012: 281).

In a special issue of *Isis*, Galison (2008) deals directly with the theme of this relationship between HS and PS. He points out ten problems where history and philosophy have to come together to find meaningful solutions. These include the challenge of defining 'context' that is sensitive to both historical and philosophical uses of this notion and defining 'purity' in the context of pure and applied science. He also discusses the increasing emphasis on the idea of the 'local'. He further argues that microhistory, in looking too closely at events, becomes a problem since the big picture might go missing; it is also the case that there are 'aspects of scientific practice that simply do not reduce to the local' (ibid. 122). Moreover, historicism makes everything historical leaving no space for philosophy to be meaningful. Friedman (2008), in the same volume, expands on the relationship between HS and PS, and points out how philosophy of science itself should be seen historically, thereby questioning a particular philosophical method which gives primacy to the author/text in an ahistorical manner. He also suggests a shift to 'tradition' as being 'central to historical inquiry'. He goes on to add that history 'should be seen, in particular, as the birth, unfolding, evolution, transformation, and perhaps (most important) mutual interaction and entanglement of a very large number of traditions constituting the extraordinarily complex and ever-changing fabric of human culture' (ibid. 133).

What is striking in these discussions is that there is no mention of the non-European traditions in all these debates about HS and PS. Even in the invocations of 'tradition' and the 'ever-changing fabric of human culture' there is no mention of the possible histories of the non-West which might be of interest to this debate. Galison refers to the 'pure and mixed mathematics of the Greeks' but it is ironic that he doesn't invoke the purely mixed nature of Indian mathematics. He makes a similar comment about purity in logic but again misses this connection to Indian logic. Yet another theme in his paper is that of scientific doubt – this too has strong resonance with Indian philosophical traditions. Even when Laudan (1992) emphasizes the shift to the local in contemporary history of science, and through it leading to the mix of sociology and politics in HS, she is silent about the extension of the idea of the local to non-Western societies. But as arguments from multicultural origins of science and connective history of science show, the very idea of the local in HS needs to be – paradoxically – broadened.

Although there have been claims about the problematical relation between history and philosophy of science, we have also seen examples where work in history of science proved to be of seminal importance in philosophy of science. These examples from history of science helped philosophers refine the nature of science and scientific methodology. So the first question that one can address to this project of CHS is this: does CHS matter to our understanding of the nature of science and scientific practice? If we are not able to establish this relationship, then there is a great danger of CHS becoming an exotic cultural history, and worse, a politically correct history in the age of globalization.

This raises an important question for CHS. Is this history explicitly cultural history? (See Dear 1995) Is it possible to even consider the possibility of a CHS without the overbearing presence of cultural history? What makes it more difficult for CHS is that the point of contention is the notion of science itself. For this history, the subject matter of science is not available for historicizing; one needs to postulate, discover and legitimize the claim that there is an object called science available for this historicizing. So, in a fundamental sense, conceptualization – of the philosophical variety – becomes an integral part of any CHS.

One of the claims which I am making here is that CHS can actually lead to a 'different' philosophy of science. I will establish this by showing how new concepts of science emerge from this history once we pay heed to alternate scientific practices and methods. There is a great danger in calling certain practices as 'alternate' science since this seems to go against the very idea of a unified science. But in what follows I will hopefully illustrate some convincing examples in support of this claim.

2   Can Different Histories of Science Lead to Different Philosophies of Science?

The first point to note here is that philosophy of science – whether in the dominant analytical tradition or the not so popular continental version – is still fundamentally dependent on Western philosophical categories and discourse. This is a matter of great puzzle: what is this intrinsic and essential relationship between philosophy and science that necessitates only particular philosophical traditions to make sense of science? One might argue that this belief that only Greek and European philosophical traditions are necessary for philosophy of science is because of the belief that science began in these cultures. There are two fundamental mistakes in this belief: one about the claim that science is unique to Ancient Greece and Europe, and two that even if this were so there is no convincing reason to believe that the philosophy needed to understand science should be the philosophy of the 'culture' which gave rise to science. As I have argued elsewhere (Sarukkai 2005a), this fallacy is compounded by the historical fact that the European tradition gradually found philosophy to be inimical to the practice of science since the rejection of metaphysics, and philosophy in general, from matters of science in Europe is an important component of the growth of modern science.

Note that the HS community has not been sympathetic to extending the possibility of a history of science which shares its roots in the non-west. For example, Ganguli (1929) published a paper in 1929 responding to Kaye's article on Indian mathematics which points out the many mistakes in Kaye's knowledge and interpretation of Indian mathematics. Srinivas (2008) points out how many contemporary writers continue with similar misreading or ignorance about Indian mathematics. Bala's (2006) influential book on multi-cultural history of science lists various examples of the uneasy if not inimical reception of non-western science as part of a larger global history of science. Little seems to have changed in the history of science community!

And when writers engage more seriously with the non-west and attempt to create a more 'global science', new pitfalls are created by invoking categories like 'indigenous science', 'local knowledge', 'vernacular science' and such. Tilley (2010) points out how the creation of these categories depend on many factors: primary among them is the globalization of science, the 'professionalization of anthropology', the colonial project and finally, the reaction from the colonized against dismissive reactions to their knowledge systems. Even when some of these scholars present the complex history of the creation of modern science by pointing to complex global phenomena, it is nevertheless the case that they are not able to take the intellectual contributions of non-western cultures more seriously. For example, while these authors discuss the creation of indigenous or vernacular science, they scarcely mention the philosophical and theoretical concepts underlying their practices. But at least historians of science have engaged with these possibilities of science and knowledge in the non-west; the philosophers of science have continued to be notoriously indifferent.

Elshakry (2010) observes that 'the contingency of the term "science" – shaped by different eras, geographies, and epistemological traditions – means that it is not always clear what historians of science are or even should be studying' (ibid., 98). She also suggests that historians of science are far more sensitized to the claims of science of the non-west. In response to the question as to whether science is a product of the west, she notes that the 'current historians don't bother to ask the question at all and would probably recoil at the antiquated Eurocentrism embedded in it' (ibid. 99). So, in this sense, historians of science have really opened up the way towards understanding the philosophical implications of other histories like CHS.

The very idea of western science is a historical product in which the non-west was also complicit. As Elshakry notes, it is only in the 19th century that such an idea begins to take shape, and paradoxically it was outside the west that this idea of western science began. She also points to the role that missionaries played in demarcating the idea of western science from the science of that culture, such as Chinese science, and in doing so, started choosing specific characteristics that mark science. However, the discourse around these terms was syncretic in nature and the boundaries between western science and the other sciences were initially more porous.

Elshakry makes the important observation that it was the creation of the new discipline of history of science that begins to propagate a global ideology of science based on universal values. This effort, beginning before the First World War, began to use a new ideology of internationalism in order to reshape the idea of science. Using notions such as Scientific Revolution, this discipline departed from the earlier syncretic model in order to frame the new global science which became synonymous with western science. In a sense then, the political ideology of trying to create a unified and liberal Europe following the trauma of the wars, as well exemplified by George Sarton, played an essential role in the creation of the specific characteristics of modern science.[1] While

---

1   See also Dear (2005) on the ideology of modern science.

this movement gets institutionalized in Europe and the US, it was also the case, as Elshakry points out, that within China these distinctions also get concretized. Thus, the creation of this category 'western science' itself catalyzed the creation of other types of science around the world (ibid. 106–7). The lesson from this analysis is this: the role of HS in creating the idea of western science leads to a fundamental conflict of interest, and thus continues ideological divides which have deeply influenced not just academic disciplines but also government policies on development and education. This ideology seems to have been strengthened ironically by developments in non-western countries, so much so that it is still quite impossible to talk of sciences in these cultures without being charged with being nationalistic, anti-modern and so on.

It is important to note that such careful work in the history of science, which exhibits its complicity in establishing these dichotomies, does not have similar parallels in philosophy of science. The immediate question after this historical reflective turn is this: what are the consequences in our understanding of the other sciences given that we know this historical trajectory? What critiques of the philosophical foundations of science are possible once this critique within history of science has been made? How do we move from this historical narrative to a rethinking of the fundamental conceptual and philosophical basis of science?

If, as enough work in HS clearly shows, colonialism and imperialism influence the very creation of the larger historical and philosophical themes associated with modern science then why is there still appreciable resistance to a critical engagement with other scientific traditions in the world? Ignoring them only continues this process of colonialism and imperialism and this is more dangerous since it is now done implicitly. Why in spite of good scholarship in HS on these matters, their influence on the study of nature of science and scientific method is still questionable? In the context of non-western scientific traditions, I believe that it is the lack of enough secondary material on alternate philosophical traditions that has impeded accessibility to these ideas. Even claims that 'each locality has the capacity to become central' (Sivasundaram 2010: 158), remain utopian if there is no theoretical/philosophical engagement on how this is possible given the deep chasm between the conceptual worlds of different cultural descriptions of science – just look at the way Indian mathematics and astronomy are written and communicated (as will be discussed below). Sivasundaram concludes by pointing out while there is hope that there will be a challenge to the idea that science is western in origin and character, historians of science will be able to effectively challenge it only by drawing on newer tools. What I am trying to do is to offer one particular tool that may help us challenge this ideology. In so doing, I am hoping to illustrate the importance of drawing on philosophical frameworks to make sense of these deep historical studies.

Firstly, even if we grant that ancient science and modern science both have Greek and/or European origins, does it necessarily imply that the philosophical tools for analysing these developments must come from that same region? Obviously this geographical fallacy has little basis and I have briefly given some reasons above. Moreover there is an insidious logic to the use of this fallacy: one might argue that to understand western science one uses only western philosophical tools as is done now since both occur from the 'same' history and culture but at the same time, one continues to use these western philosophical tools to judge non-western science.

Instead of belabouring this point, I will approach it from another direction – namely, look at how Greek and European philosophy have been used in analyzing science and through that inquire whether other philosophical traditions make sense. Consider a few important scientific concepts which also have a surplus of philosophical thinking around them: space, time, matter, causality, force, energy, description (language), logic and so on. There has been much written on these philosophical ideas which are also central to any analysis of science. However, these themes are not special to Greek or European philosophies. Indian philosophy, for example, engages deeply with all these themes. In fact, I would argue that non-western philosophies might sometimes contribute more usefully to the understanding of the complex scientific description of reality compared to the tools available in dominant western traditions. Let me give some specific examples.

Matter: There are many discussions on the nature of matter in the context of modern science. For example, there have been explicit attempts to relate quantum and relativity theories to ancient philosophical analysis of matter. Aristotle's view of matter is an influential idea in this project. For example, Suppes (1974) discusses this view in the context of contemporary physics. First of all, he begins by listing some of the essential features of matter for Aristotle. This includes its character as the substratum of change, its distinction from form and so on. Although he notes that this formulation of matter is different from the contemporary physics' account of it, he nevertheless points to the usefulness of drawing upon these ancient ideas of matter from Aristotle. After discussing other philosophical views of matter including Descartes and Kant, Suppes comes back to Aristotle's views of matter in the context of contemporary physics. He concludes by saying that Aristotle's theory of matter provides 'an excellent way of looking at the phenomena of high energy physics as well as at the macroscopic kind of phenomena Aristotle himself had available. I do not mean to suggest that we can pull any detailed wide scientific laws from

Aristotle. What is valuable in his concept is its wide applicability as a way of thinking about physical phenomena' (ibid. 47).

This approach is similar to many other such approaches that draw on ancient Greek philosophers to understand questions that are of philosophical relevance in the context of science. From looking at the enormously extensive literature in the philosophy of science on this and related matters, one would not be faulted for thinking that other than ancient Greeks no other civilization had thought about matter or had formulated any theory of matter, motion, space, time and so on. But how wrong one would be if she came to this conclusion! All that she has to do – unlike this large community of philosophers of science – is to read about theories of matter in Indian philosophical traditions. It might surprise one to realize that there is an enormous amount of intelligent philosophical discussion on matter in these traditions, some of which are similar to Aristotle's and others which radically depart from his views. For example, the discussion on substance, inherence, quality and change are basic to the metaphysics of Indian theories.[2] Moreover, there are many competing schools of philosophers who respond to these metaphysical claims of substance etc. Suppes shows how philosophy of science can draw upon an ancient theory of matter but his work is also illustrative of how one can be blind to Eurocentrism that seems to lurk at the heart of philosophy of science.

Similar comments hold for all the other themes mentioned above. For example, consider the extensive literature on motion, a theme that is central to the very idea of physics. While standard texts begin with Zeno's paradox of motion, there are similar paradoxes of motion, most notably in Nagarjuna's analysis of motion. There are also completely different frameworks for analysing motion as, for example, in Jain philosophy which begins by postulating both principles of motion and 'inertia' to explain motion.

Logic: I will consider one more illustrative example, that of logic. I choose this theme because of its close association with scientific method and also because any mention of logic within philosophy of science is exclusively limited to western logic, starting with the Greeks. There is much that can be discussed in the context of logic and science (Sarukkai 2005a), but I will limit myself here to making one comment. Indian logic for long had not been accepted as 'logic' by most western scholars.[3] One of the dominant reasons was that the description of inferences included empirical observations and was always in terms of cognitive states. Both these features meant, for modern logicians, that this analysis of inference had not moved away from psychologism. However, this

---

2 For example, see Mohanty (2000).

3 See Ganeri (2001) for a history of reception of Indian logic.

claim has been challenged on various grounds (Matilal 1985; Mohanty 1992). but of relevance to science, it is not even that important. The fact that Indian logic does not make a 'formal' distinction between the empirical and the formal is important in the context of science since scientific methodology is primarily an attempt to integrate these two. Jain logic, which is a multi-valued logic, is also a useful tool to look at other ways of understanding such logical structures. There is an intimate connection between Mill's methods and ideas of falsification in Indian logic.[4] However, philosophy of science has continued to ignore the possibilities of these engagements, although it is difficult to find any meaningful reason for this neglect.

Indeed a large number of other ideas that occur in philosophy of science have been analysed in great detail in Indian and Chinese philosophies. So the continued absence of an engagement with these philosophical ideas within philosophy of science and history of science is a source of great mystery – to put it most politely.

3 Will Connective Histories of Science Lead to New Insights on Scientific Method?

It is not news that other civilizations were engaged in activities that are scientific. Pioneering work in history of science has given us a large corpus of work on the scientific output of cultures like India and China. In the Indian case, the extensive work on Indian metallurgy, chemistry and mathematics – to give a few examples – have conclusively proved the presence of an active theoretical and practical engagement with activities that seem to be similar to other such activities in early Greece and later Europe. However, this does not mean that there was a universal way of doing and creating science since there are major differences between these practices in different cultures.

In spite of this scholarship, it is still hazardous to talk about these Indian practices as science. In an international meeting in January 2013 at Bangalore, when I spoke about some aspects of Indian mathematics such as their calculation of irrational numbers and their use of negative numbers, drawing on work by eminent Indian and non-Indian scholars, one of the members of the audience said that I was indulging in 'cultural pride'. It was ironic that almost all the talks before referred to Plato and Aristotle and not once was that seen to be a problem of cultural pride. Reference to not just Plato and Aristotle, but also to the pre-Socratics, as scientists is often accepted without a murmur

---

4 See Sarukkai (2005a) for a detailed discussion on this and related topics.

but what is Science?

but any mention of Indian or Chinese mathematicians or astronomers in the same way is seen as an expression of right-wing 'culturalism'.

I want to suggest here that one of the major reasons for this continued defensive mentality to any utterance of the idea of non-western science is primarily due to the skewed mainstream history of science which does not take into account non-western contributions in the creation of science (ironical considering the work in HS which questions this view). We need to take this ideology of the traditional history of science seriously for the harm it has created to non-western societies – the harm extends from their students to government policies and indeed has had a great impact on these cultures. An exclusivist history of science that keeps the possibility of the scientific imagination within a constructed Greek and European history does great violence not only to other non-western cultures but also to the very spirit of the scientific quest. For example, a continued point of dismissal of the ayurvedic system of medicine by mainstream doctors and the institution of medicine in India is that it is not science.

Secondly, with this restricted definition of science – not in terms of its practice but primarily in terms of the community which practices it – there is much about the nature of science that is not available for analysis and critique. If we accept the idea that science was more pervasive than usually thought, then we need to confront the different styles, concepts and methods of doing science. CHS will open up this possibility. Thus, at least for me, CHS is an important path to understand the complex practices of science, one that is not mired in metaphysical preoccupations of the Greek philosophers.

In other words, engagement with CHS would enable us to consider a larger set of methods, practices and theories as being scientific, which would then allow us to at least critique if not revise our standard understanding of science and scientific method. Since CHS by definition draws on multiple cultural contributions – including the Greeks – we perforce have to engage with a larger conceptual world, a larger philosophical corpus of concepts and ideas through which to make sense of science. This will immediately bring to history of science awareness of the concepts of matter, space and causality, for example, drawn from different philosophical traditions, thereby radically change the standard approaches in philosophy of science to some of these questions.

Even as I say this, I must note that there is much in philosophy of science that is about contemporary issues of scientific theory and practice. However, even in these cases, debates on realism and antirealism, arguments of efficacy and indispensability, the metaphysics of necessity and laws, still draw on philosophical concepts that are also available in alternate philosophical traditions. There is no reason to believe that these alternative philosophical ideas are irrelevant to such contemporary concerns of philosophy of science. I will illustrate this possibility through some specific examples.

CHS will by necessity have to deal with and incorporate alternate world-views and philosophical concepts. When a particular idea gets into circulation, a historical analysis of the same will exhibit the ways by which the philosophical presuppositions get mediated or even erased. For example, CHS can show how the concept of zero and indeterminate equations from Indian mathematics were absorbed by other cultures and what mathematical as well as metaphysical presuppositions were changed and challenged as a consequence.

4   Example of Indian Mathematics

Do the many facets of science in India matter to our understanding of the notion of science? Let me first list the examples of science made on behalf of Indian science: ayurveda, arithmetic, algebra and early ideas of calculus, invention of new technologies including steel and zinc, linguistics and study of grammar, theory of knowledge and truth, study of inference and so on. Some of these are directly related to scientific practice while others such as epistemology and logic have been the bedrock on which modern notions of science are based.

I am not interested here in the debate on whether these should be called science or not. The colonial discourse on Indian science is a good pointer to the pitfalls of this debate. When the British encountered many Indian inventions in science and technology, they made use of them in order to establish their own industries but refused to acknowledge that these processes were part of scientific rationality (Adas 1989; Alvares 1991). Claims that these Indian inventions were more a product of 'doing' rather than 'knowing', specifically a theoretical mode of knowing, made it easy for them to reject the claim of science to almost all intellectual contributions from India.

I had discussed earlier the philosophical influence of Kuhn's work on history of science. In spite of an ambiguous relationship between the disciplines of history and philosophy of science, we do have such examples of the use of history of science for broadening the discourse in philosophy of science. There are two paths through which Indian and other cultural traditions can influence and modify the present discourse in PS. One is by directly reflecting on the scientific texts of these traditions. This is quite common in PS where an analysis of scientific texts and theories offer material for philosophical discourse. The other path is by drawing on the history of Indian science – this is closer to Kuhn's example. The latter path needs a history of science in place – history gives the context, the background and the surrounding factors in which these activities took place. Such a history of science is available in parts and CHS offers us an expanded view of these specifically local histories of science. It may be useful here to distinguish between 'local histories of science' and 'local histories of local science'. A large amount of history of Indian science and technology seem to fall under the latter type. CHS is a move towards a 'global history of local science' because it introduces an important element of continuity. If we accept that certain ideas and techniques were borrowed, modified or otherwise used by other cultures, then it forces us to give an account of the original idea and its relation to the modified/borrowed one.

But merely establishing a historical narrative under the rubric of 'connective history' can only be a starting point. There are major conceptual issues that need to be overcome. First is the dispute over the idea of science. One of the primary ways by which the title of science is denied to non-western intellectual traditions is through the invocation of terms such as logic, scientific method, evidence, prediction and so on. The global histories of science have moved away from identifying science entirely with the European experience. However, there is much in that narrative which does not address the question which many people (particularly scientists) have: the mere claim that these cultures had science is not convincing since we see little commonality between what is done in the name of science today and what we want to call as science. This is true as much for Europe as it is for Asia since there is little in common between the Greek theories of science and modern science. But there is something else which is alien in the case of non-western cultures. One measure of this alien character is in the way science is described; I will use Indian mathematics to illustrate this.

There has been an increasing amount of available material on Indian mathematics but it seems to have had little influence on the traditional HPS community. The question I want to begin with is this: if we do accept these Indian practices as part of a continuous history of science then does it help us understand some contemporary issues in philosophy of science better?

I will look at three problems that are of contemporary concern in mathematics as well as in philosophy of mathematics. First is the theme of Platonism, in particular, the difficulty in understanding mathematics beyond the purview of Platonism. Second is the complex relationship between mathematics and the world, which finds expression in the fertile debate on the 'applicability of mathematics' and third is the relationship between mathematics, language and symbolism. All these three themes are not only broad but deep and have influenced not only the understanding of mathematics today but also its use in the sciences and the social sciences, as well in science education and science policy.

Platonism is the most enduring story of mathematics. Traced back to Plato's theory of forms, Platonism has become the default theory for understanding mathematical entities. This postulate that mathematical entities inhabit a world completely different in character from the spatio-temporal ones has been used for a variety of reasons: to explain the 'universal' truth of mathematical statements as well as to explain the indispensability of mathematics. The emphasis on mathematics as a world of objects displaces the potential claim that mathematical statements are one type of descriptions of the world. We need to wonder at the price one pays for insisting on an ideology of mathematical truth which has at its foundations various religious beliefs about mathematics.

But what is worthy of our attention is the absolute lack of Platonism in Indian mathematical thinking. There are no accounts of a metaphysics of numbers which demands a completely different ontological construction. On the contrary, mathematical entities are deeply grounded in the experiential world. Yet, there is a sense of mathematical truth which is present and accepted. After all, the most important contribution of zero and the decimal system came from these mathematicians. While one might tend to read a mystical metaphysics into the number zero – particularly in relation to the *sunyavada* of the Buddhists – there is little evidence of it. Staal (2010) suggests that the idea of zero itself has origins in linguistics and thus arises as a way of making sense of language/sounds. If that is the case, not only do numbers arise only in the context of the world but even the number zero. The example of the metaphysics of the square root of 2 is another such example. While for the Pythagoreans, this number was metaphysically problematical, for the Indians the challenge was merely to calculate it since numbers were seen as integral parts of doing and building (Dani 2010). Even the concept of proofs and how they were used in mathematical discourse suggests alternate possibilities of understanding mathematics (Srinivas 2008).

Is there anything that is of significance in this non-Platonic approach to the existence of numbers? Two domains that can draw a lot from this approach are the philosophy of mathematics as well as mathematics education. In the philosophy of mathematics, there have been attempts to develop anti-Platonist, including fictionalist, perspectives on mathematics.[5] However, for a large

---

5   For example, see Field (1989).

majority of scientists and mathematicians, Platonism seems to be an obvious fact of mathematics. Even the anti-Platonic views in contemporary philosophy of mathematics do not see fit to draw upon the nature of Indian and Chinese mathematics. I will give one example where a non-Platonic understanding of mathematics can be useful and relevant.

Consider the long debate on infinity and infinitesimals. Part of the problem in understanding these terms lay in their ontology. What did these terms really correspond to? In Kerala mathematics, the first ideas of calculus occur in terms of the descriptions of infinite series and approximations to these.[6] Are these questions on the meaning of infinitesimals relevant to their constructions? I do not think so. They are purely a particular linguistic way of describing some specific actions and thus are expressed in their own ways. Another example is that of negative numbers. Mumford (2010) discusses in detail how the European mathematicians for long tried to reduce −1 into a description of positive entities whereas Indian and Chinese mathematicians had no problem centuries earlier to accept these numbers in their business calculations. Such a view of negative numbers was extremely influential in the development of algebra including finding the solutions for indeterminate equations.

Secondly, this non-Platonic view is extremely relevant to the question of the applicability of mathematics in the sciences. Galileo's observation that mathematics is the 'language' of nature has been far more influential than it should have been. Rather than accept that mathematics is a specific description of nature, Galileo's claim about the nature of mathematics makes it into a special language, one that is fundamentally different from natural languages. Generations of scientists from Galileo have echoed this point in one way or the other. This list of scientists includes Newton, Einstein, Feynman and others. (Steiner 1998). Wigner coined the catchy phrase 'unreasonable effectiveness of mathematics' specifically to point to this problematical but intriguing relation between mathematics and its applicability: the mysteriousness lies in the fact that mathematics is created independent of the world and yet it seems necessary for the description of the world.

What is really so mysterious (a word used by Einstein in this context) about the use of mathematics? The major reason for this mystery is Platonism. If mathematical entities exist in a non-spatio-temporal world then how do we spatio-temporal beings have knowledge of them? For these scientists, who viewed mathematics along such a non-empirical axis, the use of mathematics was surprising. Its 'natural match' with physical concepts was a source of

---

6   See Ramasubramanian and Srinivas (2010).

mystery only if we first begin with a clear disjunction between mathematics and the world (See Sarukkai 2005b). It is precisely this point which Indian mathematics would challenge. Mathematics is essential to this world; it arises from this world and through human action. The puzzle of applicability will take on a completely different form if we begin with the assumption that mathematics is enworlded and embodied. Interestingly, this is a position that has now gained some ground through the framework of cognitive studies (See Lakoff and Nunez 2000) but in a predictable replay these approaches also make no mention of such approaches in non-western traditions.

Thirdly, this is related to a larger issue – the relation between mathematics and natural language, and between mathematics and symbolism. The symbolism that characterizes modern mathematics is comparatively new. Earlier traditions wrote mathematics differently. In particular, the writing of Indian mathematics suggests alternate ways of engaging with prose and poetic registers. Will these discursive practices be useful for the teaching of mathematics, especially since work in mathematics education often point to symbolization as one of the major obstacles in the teaching of mathematics? (See Sarukkai 2007).

As yet another example, one can consider the challenges in the acceptance of Ayurveda as a science. Even in Indian medical schools, there is a constant tension between allopathy and Ayurveda. Many allopathic doctors openly reject Ayurveda as medical science in spite of the empirical successes of its medicine. Taking Ayurveda into the fold of a global and connective history of science is not enough unless one can engage with its theoretical presuppositions, which include a completely different understanding of the nature of the body and the meaning of disease, health and cure. Merely placing it in a historical trajectory will not dissipate the suspicion that one is being more politically correct than being historically accurate. More than other histories of sciences including local and global histories, it is connective history of science which should and can influence philosophy of science as well draw meaningfully from it.

To do any of the above, even the connective history of science has to perforce draw upon philosophy and philosophy of science. It does not make much meaning to refer to other cultural practices as 'science' but yet not engage with the theoretical and philosophical discourse available in those cultures. Given that the underlying metaphysics of these non-western traditions of science and mathematics is quite different from that of the dominant paradigms of the west, it is necessary to decode their philosophies while writing down their histories.

Bibliography

Adas, M. (1989). *Machines as the measure of men*. Ithaca: Cornell University Press.

Alvares, C. (1991). *Decolonising history*. New York: The Apex Press.

Ariew, R. and B. Peter (1986). "Duhem on Maxwell: A Case-Study in the Interrelations of History of Science and Philosophy of Science," *PSA: Proceedings of the Biennial Meeting of the Philosophy of Science Association*. Volume 1, pp. 145–156.

Bala, A. (2006). *The Dialogue of Civilizations in the Birth of Modern Science*. New York: Palgrave Macmillan.

Burian, R.M. "More than a Marriage of Convenience: On the Inextricability of History and Philosophy of Science," *Philosophy of Science* (1977) 44 (1): 1–42.

Dani, S.G. "Geometry in the Sulvasutras," in C.S. Seshadri (ed.) *Studies in the History of Indian Mathematics*. New Delhi: Hindusthan Book Agency. (2010: 9–37)

Dear, P. "Cultural History of Science: An Overview with Reflections," *Science, Technology, & Human Values* (1995) 20 (2): 150–170.

———— "What Is the History of Science the History Of? Early Modern Roots of the Ideology of Modern Science," *Isis* (2005) 96 (3): 390–406.

Elshakry, M. "When Science Became Western: Historiographical Reflections," *Isis* (2010) 101(1): 98–109.

Field, H. (1989). *Realism, Mathematics & Modality*. Oxford: Basil Blackwell.

Friedman, M. "History and Philosophy of Science in a New Key," *Isis* (2008) 99(1): 125–134.

Galison, P. "Ten Problems in History and Philosophy of Science," *Isis* (2008) 99(1): 111–124.

Ganeri, J. (ed.) (2001). *Indian Logic: A Reader*. Surrey: Curzon Press.

Ganguli, S. "Notes on Indian Mathematics. A Criticism of George Rusby Kaye's Interpretation," *Isis* (1929) 12(1): 132–145.

Hanson, N.R. "The Irrelevance of History of Science to Philosophy of Science," *The Journal of Philosophy* (1962) 59(21): 574–586.

Harrison, M. "Science and the British Empire," *Isis* (2005) 96(1): 56–63.

Hoyningen-Huene, P. "On Thomas Kuhn's Philosophical Significance," *Configurations* (1998) 6(1): 1–14.

———— "Philosophical Elements in Thomas Kuhn's Historiography of Science," *Theoria* (2012) 27(3): 281–292.

Lakoff, G. and R.E. Nunez (2000). *Where Mathematics Come From: How The Embodied Mind Brings Mathematics Into Being*. New York: Basic Books.

Laudan, R. "The 'New' History of Science: Implications for Philosophy of Science," in *PSA: Proceedings of the Biennial Meeting of the Philosophy of Science Association 1992, Volume Two: Symposia and Invited Papers* (1992), pp. 476–481.

Matilal, B.K. (1985). *Logic, language and reality: Indian philosophy and contemporary issues*. Delhi: Motilal Banarsidass.

Mohanty, J.N. (1992). *Reason and tradition in Indian thought: An essay on the nature of Indian philosophical thinking*. Oxford: Clarendon Press.

——— (2000). *Classical Indian philosophy*. Lanham: Rowman & Littlefield.

Mumford, D. "What's so Baffling about Negative Numbers? A Cross-Cultural Comparison," in C.S. Seshadri (ed.) *Studies in the History of Indian Mathematics*. New Delhi: Hindusthan Book Agency. (2010: 113–143)

Nickles, T. "Philosophy of Science and History of Science," *Osiris* (1995) 10: 138–163.

Pinnick, C. and G. Gale. "Philosophy of Science and History of Science: A Troubling Interaction," *Journal for General Philosophy of Science* (2000) 31(1): 109–125.

Ramasbramanian, K. and M.D. Srinivas "Development of Calculus in India," in C.S. Seshadri (ed.) *Studies in the History of Indian Mathematics*. New Delhi: Hindusthan Book Agency. (2010: 201–286)

Sarukkai, S. (2005a). *Indian Philosophy and Philosophy of Science*. New Delhi: Motilal Banarsidass.

——— (2005b). "Revisiting the 'Unreasonable Effectiveness' of Mathematics," *Current Science* (2005) 88(3): 415–423.

——— (2007). "Philosophy of Science and its Implications for Science and Mathematics Education," in Choksi, B. and Natarajan, C. (eds.) *The epiSTEME Reviews Vol. 2, Research Trends in Science, Technology and Mathematics Education*, New Delhi: Macmillan India.

Sivasundaram, S. "Sciences and the Global: On Methods, Questions, and Theory," *Isis* (2010) 101(1): 146–158.

Srinivas, M.D. (2008). "Proofs in Indian Mathematics," in H. Selin (ed.) *Encyclopaedia of the History of Science, Technology, and Medicine in Non-Western Cultures*. Springer.

Staal, F. (2010). "On the Origins of Zero," in C.S. Seshadri (ed.) *Studies in the History of Indian Mathematics*. New Delhi: Hindusthan Book Agency.

Steiner, M. (1998). *The Applicability of Mathematics as a Philosophical Problem*. Cambridge: Harvard University Press.

Suppes, P. "Aristotle's Concept of Matter and Its Relation to Modern Concepts of Matter," *Synthese* (1974) 28(1): 27–50.

Tilley, H. "Global Histories, Vernacular Science, and African Genealogies; or, Is the History of Science Ready for the World?" *Isis* (2010) 101(1): 110–119.

CHAPTER 3

Kuhn, Nisbett, Thought Experiments, and the Needham Question

James Robert Brown

1   Introduction

Why was China surpassed scientifically by the West?, or, to put it another way, Why did Europe, rather than China, have a scientific revolution, given that China was so well prepared for it? The Needham Question, as it is widely know, has prompted many different answers and extensive and on-going scholarly debates. There is a whole family of related questions: Why did the West rather than China have an industrial revolution? Why did the West rather than China adopt capitalism as an economic system?, and so on.

On hearing the question, some might think it contains racist overtones and smacks of European smugness. Such a charge could not be laid at Needham's feet; he was a great admirer of China, past and present. Chinese scholars and the government of China have gladly embraced the question and promoted the search for answers. Others have wondered why China? Can't similar questions be asked about India or Egypt, and so on? Perhaps, but the Needham question about China is based on the similar situations in China and Europe in the late medieval and early modern period, a similarity that was not present in other countries. This cluster of questions prompts many others. For instance, it raises what we might call inverse Needham questions, such as "Why was China advancing scientifically in the middle ages when Europe, with its rich Greek and Roman heritage, was stagnating?" Interesting though such questions are, I will ignore variations on the Needham theme and stick to the central question.

I might point out that even those not inspired by Needham's work are nevertheless interested in the same issue. The British historian Niall Ferguson has no particular interest in the Needham question, but poses a clearly related challenge when he writes:

> Western predominance was a historical reality after around 1500, and certainly after 1800. In that year, Europe and its new world offshoots accounted for 12 per cent of the world's population and (already) around 27 percent of its total income. By 1913, however, it was 20 per cent of the

world's population and more than half – 51 per cent – of the income. . . . Like it or not, the fact is that after 1500 the world became more Eurocentric. And understanding why that happened is the modern historian's biggest challenge. (Ferguson 2010)

Ferguson and Needham are asking the same question, but they seek very different answers. Ferguson is a Western imperialist – quite literally. He is a staunch defender of the old British Empire and an enthusiastic advocate of the new American Empire. He will seek an answer to the question of Western ascendency in the West itself, that is, in terms of what he takes to be Western virtues. Needham, by contrast, looked to features of China that somehow held it back. Although my sympathies are much more with Needham than with Ferguson when approaching this cluster of questions, it will be wise to keep the full range of possible answers in mind.

The connections among these various Needham questions (and inverse Needham questions), are presumably the obvious relations among pure science, technology, industry, and systems of political and economic organization of society. Needham, who was, of course, a Marxist, would have taken those relations for granted. It is not entirely clear which is cause and which effect, but I suspect that if we could answer any one of the questions, we would have at least some of the ingredients needed to answer the others.

My focus, however, will be on the pure science end of this spectrum of questions, not the industry-economics pole, though overlap is inevitable. So, the particular Needham question I want to address is the one most commonly asked: Why did modern science develop in the West rather than in China?

2   Proposed Answers

There have been numerous answers proposed over the years. Needham's own was to say that China was a bureaucratic state, run by its famous civil service. Such a highly centralized organization did not readily transform into the kind of bourgeois state to which medieval European feudalism gave rise. This puts the explanation in terms of social and political factors. Leibniz and Russell (who answered the question before Needham asked it), took a very different view; they thought that a lack of mathematics in China was the reason. This is to place the explanation within the intellectual realm itself. A political explanation, different from Needham's, has been proposed by David Hume and

others. They hold that competition among rival European states stimulated ✓
innovation. A united China, by contrast, lacked that sort of incentive.

Yet another kind of explanation is psychological. Kaiping Peng and Richard
Nisbett (2001) think the answer to Needham's question has much to do with ✓
very different ways of thinking – Westerners are analytic-minded, whereas the
Chinese (and Asians generally) are holistic and dialectical. I will focus on the
Peng-Nisbett account, especially as it pertains to pure science.

Ignoring money incentives + The bourgeois

3   Kuhn's Incommensurability

Attempts to answer the Needham question have important presuppositions
that have not been fully explored. Historiographic and philosophical work of
the longitudinal sort that involve changes of belief within the same culture
have raised several conceptual problems that might carry over to comparative
studies. Thomas Kuhn, for instance, claims that people who hold different paradigms, even though they are part of a single culture, "live in different worlds"
and talk past one another when discussing their rival views. If Kuhn is right,
what implications might this have when we try to compare Western science
with Chinese? The relevance for the Needham question should be obvious. It
is especially relevant when we consider whether people truly think differently,
as Nisbett believes.

To address these issues let us consider the aims of comparative history of
science. One obvious response is that it has the same aim as non-comparative
history of science – namely, historical understanding; we want to know what
actually happened and why. If we stress *comparative*, we can give a second reason: Our understanding is considerably enriched by the contrast. Comparative
history is like comparative literature or comparative science. Our grasp of
quantum mechanics is deepened by solving problems with it, but it is further
deepened by seeing how Newtonian mechanics would deal differently with the
same problems. There is also a different kind of answer to the question, a wide-
ranging response that involves a cluster of social and political concerns. Some
investigators are looking for bragging rights or articulating grievances: "Science
is a glorious Western creation" or "We discovered it first; the West stole it from
us." There is also the very different "Three cheers for multiculturalism" outlook, which is sometimes caricatured as wanting to award credit to everyone
for everything. And, of course, there are those who modestly seek something in
between, namely, the sense of intellectual justice – not to mention intellectual
pleasure – that would come from learning what really happened.

Contrasts and comparisons require mutual intelligibility. If we are to compare Newtonian physics with quantum mechanics, then we must understand them both. Similarly, if we are to contrast Western science with Chinese science, then we must find them both intelligible. This, as anyone who has ventured into the history of science knows, raises the problem of incommensurability.

Comparing things is a commonplace activity. We could compare chicken *tikka* in an upmarket Indian restaurant with McDonald's Chicken McNuggets. Of course, such a comparison may sound silly. It would be better to compare it with, say, *coq au vin* from a good French bistro. After all, it seems only fair to compare one of Asia's best with one of the best the West can offer, which is unlikely to come from a burger chain.

One of the important facts about cooking is that it is largely non-propositional. We can learn how to make *coq au vin*, without knowing a word of French, by watching how it is done. And we can similarly learn how to make chicken *tikka* without sharing a common language with an Indian chef. After the initial lesson, we practice, much like learning to ride a bike. I might also add that in spite of cultural differences – often considerable – it does not take long to appreciate the cuisines of other cultures and to do a fair job of replicating them.

However, we are not here concerned with cooking or any other branch of technology, but with the more theoretical side of scientific ideas and their history. If there is a problem with incommensurability, then here is where it arises. The term comes from Thomas Kuhn and Paul Feyerabend who were concerned with the history of Western science and wondered how words in distinct theories could maintain their meaning across scientific revolutions. How, for instance, could the concept of "mass," as used by Newton, mean the same thing as "mass" used by Einstein? They wanted to explain persistent disagreements in the history of science that defied normal explanation. For instance, if rivals were aware of different experimental results, then that would do nicely as an explanation for continued disagreement, but such explanations are often lacking. It seemed to Kuhn and Feyerabend that rival scientists were often talking past one another, because they seemed to mean different things even when they used the same words.

Though somewhat vague, this idea fits nicely with the theory of language that both Kuhn and Feyerabend held. Paradoxically, they inherited it from the Positivists and Logical Empiricists, the very people whose views they set out to reject. The doctrine is that theoretical terms in a theory pick up their meaning contextually or implicitly by occurring in statements made by the theory. Thus, "mass" gets its meaning (at least in part) by appearing in Newton's second law: *Force = mass × acceleration*. The consequence of this doctrine is obvious: If we change the basic laws, then the meaning of terms such as "mass" will also change.[1]

The doctrine of incommensurability arose from thinking about the history of Western science, but it obviously carries over to science in different cultures. If we have trouble understanding our own ancestors, we should certainly have as much or more trouble with other cultures, both in their contemporary views and perhaps even more with their past theories. It might also carry over to cultural differences that transcend particular theories, to styles of thinking, for instance. This is something we will take up later.

There are types and degrees of incommensurability. Originally Kuhn and Feyerabend were rather modest in their claims. They chose the term with an eye to Greek mathematics, specifically the incommensurability of the diagonal with the side of a square. The unit square has sides of length 1 and diagonal of length $\sqrt{2}$. However, there are no whole numbers p and q, such that $\sqrt{2} = p/q$. In other words, it is not a rational number. If one were limited to the natural numbers, 0, 1, 2, . . ., and ratios of them (the rational numbers), and the usual operations of addition, subtraction, multiplication, division, and taking roots, then one could not express the number $\sqrt{2}$. That is what is meant by saying the diagonal is incommensurable with the side.

The analogy with mathematics expresses the dramatic aspect of the philosophical doctrine. But Kuhn and Feyerabend were also aware of Eudoxus's theory of exhaustion, a forerunner of infinite series. Even though there are no whole numbers p and q such that p/q exactly equals $\sqrt{2}$, we can approximate it as close as we might wish. Thus, the sequence 14/10, 141/100, 1414/1000, 14142/10000, 141421/100000, . . . gets closer and closer to $\sqrt{2}$. This would seem to lessen the dramatic aspects of incommensurability to the point of insignificance, since we could get as close as we like, or need, to the precise meaning of a term.

There is, however, a stronger version of the doctrine, and it is the possibility of this stronger version that we must consider. Terms in different theories mean different things and there is nothing in the way of a bridge between them, not even approximately. The strong version of incommensurability is about more than word meaning. There is also incommensurability of

---

1 Kuhn and Feyerabend were not the first to notice this. The mathematician David Hilbert had an extensive debate with the logician and philosopher Gottlob Frege on this very point. Hilbert claimed that geometrical terms such as "point" are defined by the geometric axioms. Frege objected that if we switch from Euclidean to non-Euclidean geometry, we would change the meaning of the word "point". Frege took this to be absurd, but Hilbert was happy to accept the consequence. For a discussion, see Brown (2008).

methods, standards of evaluations, and much more. What we observe depends on what we already believe and expect; this is the theory-ladenness of observation. What counts as a problem and as a solution to it, and what are taken to be legitimate ways of calculating and measuring are tied to a particular theory, or paradigm, to use Kuhn's famous term. It all comes as a package. In the strong version, people "live in different worlds," as Kuhn infamously put it. They have nothing in common and mutual understanding is hopeless. This is the strong version of incommensurability we must worry about when doing cross-cultural history of science.

Two important questions arise. Which aspects of science are we comparing? And what do we mean by "science," anyway? In addressing the latter question, we need to be quite liberal and should define science only to the extent needed to get on with our task. Science is a cluster concept, usually involving many ingredients including the following: it aims to explain phenomena of interest and to provide an understanding of them; it aims to predict and control, possibly for human benefit; and it amuses and entertains us, while trying to satisfy our curiosity.

If we gave a strict Popper-style definition of science we would likely rule out several relevant things. Popper insisted on a theory being falsifiable in principle in order to be considered a part of legitimate science. This would be a terrible policy when it comes to the history of science, an inherently messy business. After all, many theories, East and West, were not falsifiable in Popper's sense but are still historically important. We should also not endorse a sharp distinction between the natural and the social sciences, or between science and technology. At the very least we should not be too insistent on our own current standard distinctions. Western investigators sometimes do this, looking, say, for so-called indigenous medicines that seem to be efficacious. They want the beneficial practices but reject the indigenous theoretical understanding of it. The theory/practice distinction might be quite different for indigenous people. Imposing a different framework on the indigenous account could well lead to a serious misunderstanding of the actual beliefs of indigenous scientists. I am not asserting that the Western pure science/technology distinction is wrong or even that it cannot be profitably imposed on, say, Chinese medicine. The mistake comes in thinking that the same distinction holds when viewed through Chinese eyes. That is something that would need to be established, not taken for granted. It is an important aspect of the general problem of incommensurability.

As for what we are comparing in any cross-cultural study, there are several possibilities. We could compare theories. This, in principle, would seem to be similar to comparisons within a single culture, such as the comparison of

Lamarck's account of evolution with Darwin's. It could be narrower (in some sense) than this, for instance, comparing achievements in Greek and Indian mathematics concerning a single result, such as the Pythagorean theorem. Or it could be very much broader, comparing general outlooks, paradigms in the sense of distinct *weltanschauung*, for instance, a mechanical versus a teleological view of nature.

However, I would like to draw attention to a relatively high level of theory and methodology that might be called *styles of thinking*.[2]

4 Nisbett and Peng and Styles of Thinking

It has been suggested that Asians and Europeans think differently, and it has even been suggested that this is a key to answering the Needham question. Richard Nisbett and his co-author Kaiping Peng have made both of these claims.

First, they express the claim about thinking styles as follows:

> We ... are pursuing the general notion that East Asians influenced by Chinese cultural tradition are cognitively integral and holistic, attending to the perceptual and cognitive field as a whole. In contrast, Westerners are prone to differentiate the object from the field and to reason analytically about its behavior, categorizing it and using rules about categories to understand its behavior. ... There is now considerable evidence, for example, that causal attribution differs across cultures, with Asians being inclined to attribute to context the sorts of actions that Westerners attribute to dispositions of the object ... (Peng and Nisbett 1999: 750)

> The tendency toward dialecticism of Easterners may thus be seen as part of a general system of thought in which attention is directed outward toward the environment, and complexity and change and contradiction are therefore salient. The Western tendency toward logical reasoning may be seen as due to a focus on the object, with its presumably fixed attributes, resulting in a general system of thought in which rules and categories concerning the object are viewed as essential. (Peng and Nisbett 1999: 750–751)

---

2 Ian Hacking has written a great deal of important work on styles of thinking. This, however, is not what I have in mind here, though there may be overlap.

Next, Peng and Nisbett express the consequence, as they see it, for the Needham question, or Needham's paradox, as they call it:

> Our contention that East Asians are inclined toward holism and dialecticism, whereas Westerners are inclined toward analytic thought focusing on the object may be helpful in resolving "Needham's paradox" ... One of the factors contributing to Needham's paradox could be naïve dialecticism. By emphasizing change, contradiction, and covariation, naive dialecticism restricts any reductive, analytic, and logical quest for understanding nature and the world. (Peng and Nisbett 1999: 751)

Peng and Nisbett stress what they call dialectical reasoning. In saying what it means, they cite Karl Marx and mention that it involves the transition: thesis, antithesis, synthesis. It is somewhat paradoxical that Marx, a quintessential Western Enlightenment thinker, should be the exemplar of this, but we will let that pass. The Peng-Nisbett distinction between different styles is best understood by looking at their examples. I will examine just one, Galileo's thought experiment to show all bodies fall at the same rate.

Let me emphasize what is at issue. If A and B initially disagree on some proposition P, but discover that they agree on everything else, then they can very likely settle the question of P. One of them will produce the relevant evidence, convincing the other to shift his or her belief. However, when A and B disagree about P and also come from different paradigms, then, given the doctrine of incommensurability, they would seem to mean different things by P. Coming to an agreement on P will only happen when one of them switches paradigms to that held by the other.

It gets worse. Ordinary incommensurability concerning the meaning of terms is problem enough, but it is seriously exacerbated by Peng and Nisbett when they add what they take to be a very deep and enduring psychological dimension, namely, an analytic vs dialectical difference in styles of thinking. They take this difference to be deep and long-standing. They also take this difference to be cultural; there is no suggestion of a genetic origin, and yet, they say, it has been present for "thousands of years." It won't be easily shaken off.

If they are right, then Western and Chinese paradigms will be fundamentally different. Perhaps – but this would be an arguable point – the difference between any pair of Western paradigms is less than the difference between any Western paradigm and any Chinese paradigm. If they are right about this, then as well as the usual range of issues involving incommensurability, we need to consider an even wider set of problems. Not only would we have trouble evaluating rival paradigms using theory-laden observations, but we

[handwritten margin note: Not just Nisbett — Nakamura etc. + Granet language etc.]

must evaluate rival paradigms using very different standards of reasoning. This is the strongest form of incommensurability imaginable. Even for Kuhn, an explicit contradiction was seen as a problem whether one was inside a particular paradigm or outside. But for Peng and Nisbett, taking contradictions seriously is up for grabs.

Kuhn eventually backed away from his more controversial doctrines and allowed that many of the problems of incommensurability could be overcome. In a famous essay, Kuhn (1977) proposed five criteria that transcend any paradigm. Paradigms could be evaluated on grounds of (1) accuracy, (2) consistency (internal and with other theories), (3) scope, (4) simplicity, and (5) fruitfulness. This requires, as a cognitive minimum, some basic ideas about scientific reasoning that are common to all humans at all times. If Peng and Nisbett are right about the cultural depth of analytic vs dialectic reasoning, then Kuhn's list of transcendent principles is not open to all of humanity. There will be even fewer grounds for comparison and we cannot tell who is and who is not making progress. If this turned out to be the case, then we could not even pose the Needham question, much less answer it, since we could not say that as a matter of fact the West leaped ahead of China.

The problem is not with analytic and dialectic thinking *per se*. They could, after all, be paradigm bound. Holistic thinking, for instance, is part of Aristotelian physics just as reductionistic thinking is natural to mechanistic atomism. But as long as we have some paradigm transcendent principles, then we can evaluate holism or reductionism along with the rest of the paradigm. If they are outside all theorizing, then we are in a hopeless mess. I shall argue, however, that Peng and Nisbett are wrong in their outlook. There is no such deep cultural difference. When we encounter these different styles of thinking, we should attribute them to rival paradigms, not to deep cultural factors that cannot be overturned by reason and evidence, that is, overturned by more and better science.

[handwritten margin note: language effect ignored; deep culture effect — conformism fatalism Collins]

5   Galileo's Thought Experiment

A thought experiment by Galileo is used by Peng and Nisbett to illustrate their point. Galileo famously argued that all bodies fall at the same rate, regardless of their weight. Peng and Nisbett presented two versions of this famous thought experiment to groups of students studying at their university in the USA. The student subjects were born and raised either in the USA or in Taiwan. None of them, unfortunately, were physics students, though some were students of other sciences. They were asked to evaluate the two versions of the thought experiment for "persuasiveness" and for "liking." I suppose that "persuasive-ness" is readily understood, though it admits degrees. Let us assume that it crosses some threshold and constitutes *belief*, that is, to say something is per-suasive means that we believe it is true. "Liking," however, is much more vague but could nevertheless be a useful and objective concept because there may be arguments that strike us as wrong and unpersuasive and yet be remarkably attractive. The ontological argument for the existence of God, for instance, has always struck me as ingenious, even though I think it fails. Keeping this distinc-tion between "persuasive" and "liking" in mind, let us look closer at the Galileo example, as Peng and Nisbett present it.

The first version of Galileo's thought experiment that they present is close to Galileo's original; it is in their words.

Galileo's Argument Against Aristotle's Assumption

Aristotle believed that the heavier a body is, the faster it falls to the ground. However, such an assumption might be false. Suppose that we have two bodies, a heavy one called H and a light one called L. Under Aristotle's assumption H will fall faster than L. Now suppose that H and L are joined together, with H on top of L. Now what happens? Well, L + H is heavier than H so by the initial assumption it should fall faster than H alone. But in the joined body L + H, L and H will each tend to fall just as fast as before they were joined, so L will act as a "brake" on H and L + H will fall slower than H alone. Hence it follows from the initial assumption that L + H will fall both faster and slower than H alone. Since this is absurd the initial assumption must be false. (Peng and Nisbett 1999: 753)

Peng and Nisbett take this to be typical Western analytic reasoning. Setting aside "Western" as possibly irrelevant, I am inclined to agree. If I were to dis-agree, it would be only to say that it is untypically brilliant. Galileo was argu-ably the greatest thought experimenter of all time and this, I think, was his best example. In passing, I should note that what Nisbett and Peng present is merely the first part of the thought experiment. After seeing the paradoxi-cal nature of Aristotle's account, we quickly recognize that the right theory of falling bodies is that *they all fall at the same rate*, regardless of how heavy they might be.

Next Nisbett and Peng construct a dialectical version of the thought experiment.

Dialectical Argument Against Aristotle's Assumption

Aristotle believed that the heavier a body is, the faster it falls to the ground. However, such an assumption might be false because this assumption is based on a belief that the physical object is free from any influences of other contextual factors ("perfect condition"), which is impossible in reality. Suppose that we have two bodies, a heavy one called H and a light one called L. If we put two of them in two different conditions, such as H in windy weather (W) and L in quiet weather (Q), now what happens? Well, the weights of the body, H or L, would not make them fall fast or slow. Instead, the weather conditions, W or Q, would make a difference. Since these kinds of contextual influences always exist, we conclude that the initial assumption must be false. (Peng and Nisbett 1999: 753)

Peng and Nisbett conducted a survey of Chinese and American students to determine whether they found the analytical or dialectical argument more persuasive. The results they found are as follows: Approximately 60% of the Chinese students found the dialectical argument more "persuasive," while only 1/3 of the American students agreed. The American students strongly preferred the analytic version. The figures for "liking" these arguments were similar.

Unable to resist the opportunity, I tried a version of the Nisbett and Peng questionnaire on the participants at the Singapore conference where this essay was first presented.[3] The result was quite different from Nisbett and Peng's. A significant majority (two thirds), who, by the way, were mostly Asian and educated in Asia, favoured the original Galileo version of the thought experiment. It mattered slightly, but not significantly, that those surveyed had studied Newtonian mechanics, at least at a high school level. Of course, the results must be taken with a grain of salt, since the conditions were far from ideal for such a test: the sample was small, most of the Asians tested had, at least in part, a western-style scientific education, and so on. However, it is grounds for a healthy degree of scepticism concerning Peng and Nisbett's results.

In his book on this topic, *The Geography of Thought: How Asians and Westerners Think Differently . . . and Why* (2003), Nisbett reports in autobiographical fashion that he has abandoned his long-held universalist views, because of this survey result and similar findings. *Universalism* is the doctrine that all people at all times think the same way. He writes that his research, especially that done with Peng, has led him:

---

3 The questionnaire presented both Galileo's argument and the dialectical argument against Aristotle's assumption followed by five questions: (i) Do you find either version persuasive? (ii) Which is more persuasive? (iii) Do you like this sort of argument? (iv) Were you raised in Asia? (v) Have you ever studied Newtonian physics?

...to the conviction that two utterly different approaches to the world have maintained themselves for thousands of years. These approaches include profoundly different social relations, views about the nature of the world, and characteristic thought processes. Each of these orientations – the Western and the Eastern – is a self-reinforcing, homeostatic system. The social practices promote the worldviews; the worldviews dictate the appropriate thought processes; and the thought processes both justify the worldviews and the social practices. (Nisbett 2003: xx)

Peng and Nisbett acknowledge two problems with their study. First, none of the subjects were students of physics. Chinese physics students might have very different intuitions about such matters and might, for all we know, have universally endorsed the analytic version of Galileo's thought experiment. A subgroup within any culture may well have very different intuitions as well as very different considered beliefs from typical members of that culture. For instance, American scientists are overwhelmingly non-religious, even though they live in one of the most religious societies in the world. When they look at bees pollinating flowers American biologists do not see the work of an intelligent designer; they see evolutionary adaptation. Developments in science and technology are almost wholly dependent on this subgroup, so it matters little how most people in that culture think.

The second problem Peng and Nisbett acknowledge is that the Chinese subjects are studying in the West. So they very likely came with or subsequently adopted Western attitudes, at least to some degree. This problem is less serious than the first, since acclimatization would have a tendency to wipe out differences. Thus, the differences that allegedly do show up, namely, a preference for dialectical arguments, can be considered rather robust, a point in favour of the Peng and Nisbett claim.

There are more serious problems to consider than the two that Peng and Nisbett raise for themselves. In thinking about them we need to keep two things in mind: their claim that Asians and Westerners tend to reason differently, and second, that this answers (in part) the Needham question.

Nisbett asserts a view reminiscent of Kuhn. "In learning a paradigm," Kuhn famously claims, "the scientist acquires theory, methods, and standards together, usually in an *inextricable* mix." (Kuhn 1970: 108) The holistic messages of Kuhn and Nisbett are not the same, but they are importantly similar. I repeat the Nisbett quote: "Each of these orientations – the Western and the Eastern – is a self-reinforcing, homeostatic system. The social practices promote the worldviews; the worldviews dictate the appropriate thought processes; and the thought processes both justify the worldviews and support the social practices." (Nisbett 2003: xx)

Always ignore The wider context

The difference is that Nisbett thinks Chinese forms of thinking have been stable for thousands of years, whereas Kuhn thinks paradigms come and go, some having a relatively short life. Holism is the common element. But we must take care not to confuse two levels of holism. Both Kuhn and Nisbett hold that there is something that we often call a worldview or *weltanschauung* that plays a guiding role, knitting several elements into an "inextricable mix." Holism at this level should not be confused with holism at an internal level. Inside their worldview, Westerners might be analytically minded, not holistic at all, while Chinese are very holistic. It is odd, but not paradoxical, to say with Kuhn that paradigms come as a package, even though one of the elements of that package is a methodology of doing science that pursues reduction and shuns context.

Nisbett's principle claim, by contrast, is quite anti-Kuhnian. Whereas Kuhn thinks that a paradigm that dictates an analytic and reductionistic methodology could be replaced by another paradigm that does not, Nisbett implicitly holds that every Western paradigm is methodologically analytic and reductionistic, and every Chinese paradigm is methodologically dialectical and holistic. This is the inevitable upshot of his account. But is this true?

Aristotelian physics, which came before the Scientific Revolution, and quantum mechanics, which came after, would seem to be serious counter-examples to this outlook. Aristotle took the world to be a kind of organism. We understand the parts in terms of the whole, just as we understand the heart in terms of its function in the body. This is classic holistic thinking.

By contrast, the spirit of much science since the Scientific Revolution is indeed reductionistic; we understand the whole in terms of the prior understanding of the parts. Classical physics takes the world to consist of isolated entities and processes. Quantum theory, however, disputes this. Quantum properties tend to be relational. That is, objects in the micro-world have their properties in relation to macroscopic measuring devices. An electron, for instance, has a position only in relation to a measuring device that can measure its position. In such a case, where its position is precisely determined, according to Heisenberg's uncertainty principle it does not have a momentum at all. A momentum measuring device could create a momentum for the electron, but it does so at the expense of giving up the property of position. Though not everyone would agree with this, such a view is widely accepted by contemporary physicists and philosophers of physics.

Relational properties are not entirely new. There are some in classical physics, weight, for instance. Mass is an intrinsic property of an object, but weight is something it has only in relation to another gravitating body, say, the earth or the moon. Quantum properties seem all to have this relational feature, which is one of the more striking characteristics of the theory. Quantum holism, as it is known, is intimately connected to a relational account of properties. In the Einstein-Podolsky-Rosen thought experiment, distant photons that were once coupled but are now moving apart are connected somehow, so that a measurement performed on one photon seems to affect the other photon, even though the influence must travel much faster than the speed of light. Quantum holism is one of the most striking and perplexing features of modern physics. This sort of holism is being exploited in quantum computing and quantum cryptography.

There is a partial reply open to Peng and Nisbett. Classical mechanics is paradigmatic of the Scientific Revolution, so it is open to them to say this sort of thinking is very un-Chinese and that is why China didn't have a counterpart to the Scientific Revolution. I say it is a partial answer, because the Aristotelian and quantum examples show that analytic, reductionistic thinking is not essential to the West; so, it undermines their claim that dialectical and holistic thinking is characteristic only of the Chinese. Still, it might provide some insight into why there was no mechanistic, reductionistic, or analytic phase in the history of Chinese science. If we think that such a phase was necessary for the Scientific Revolution, then we have an answer of sorts to the Needham question. This is not plausible, however, since the alleged rigidity of Eastern and Western thinking patterns seems overturned by the Aristotelian and quantum mechanics examples. It must also be added that contemporary physicists who are trained in holistic quantum theory nevertheless embrace the analytic version of Galileo's thought experiment.

Let us now return to the example I started with – Galileo's thought experiment. The thought experiment can be – and usually is – presented in a very neat and tidy way. Some of us find it stunningly beautiful and utterly persuasive. No doubt, this is how Galileo would like to have it received and then move on to other matters. But Galileo knew that readers in his day would have many diverse concerns. In the *Dialogo*, the character Simplicio upholds the Aristotelian outlook. He does not accept Salviati's (Galileo) conclusion that all bodies fall at the same rate, but instead challenges some of the assumptions of the thought experiment. And he is quite right to do so. What, for instance, is a body? Are joined objects one body or two? It makes a difference. As a matter of empirical fact, heavy bodies do fall faster in the atmosphere than light bodies. Galileo appeals to a vacuum, but Simplicio says this is illegitimate, since, he claims, a vacuum is impossible.[4]

Although Nisbett repeatedly claims that Westerners are more logically rigid than Asians, the evidence he produces for this is at time spurious and he seems

---

4   For detailed discussion of his arguments see Galileo, *Dialogo* pp. 61–64.

to miss occasional evidence to the contrary. Western scientists, as he rightly notes, do not like contradictions, but they do not invariably reject contradictory theories on that account. Most well-developed theories have conceptual problems that need to be fixed, but often it is quite unclear how to do this. For example, quantum electrodynamics faces an outright absurdity. We calculate the probability of various processes using a so-called perturbation series. This series is divergent. Fortunately, it is asymptotic, so we can calculate using the first few terms and get reasonable answers. Physicists realized the theory as it stands is absurd, but no one wanted to throw out a theory that otherwise could do so much. Physicists worked around the problems, constantly trying to repair the theory so that the contradiction could be contained, if not eliminated. The same could be said about a branch of mathematics, set theory, after Bertrand Russell discovered his famous paradox. Mathematicians needed set theory for working elsewhere in mathematics, so there was no question of tossing it out. But they lived with the contradiction for a few years until Zermelo found a way that seems to solve the problem.

If Nisbett misses contradictions where they exist, he sees them where they do not. In one of his tests given to Americans and Asians, he takes a pair of sentences to be contradictory. But this depends on how one reads the passage. Nisbett writes:

> The pair of statements below was typical of the more obviously contradictory ones.
>
> Statement A: A survey found that older inmates are more likely to be ones who are serving long sentences because they have committed severely violent crimes. The authors concluded that they should be held in prison even in the case of a prison population crisis.
>
> Statement B: A report on the prison overcrowding issue suggests that older inmates are less likely to commit new crimes. Therefore, if there is a prison population crisis, they should be released first. (Nisbett 2003: 181)

These are perfectly consistent, as a moment's reflection will show. It follows from (B) that the young are more likely to commit crimes. Thus, the older inmates of (A) likely committed their crimes when they were young, and, consequently, they have been in jail for quite a while. Assume that they are now considerably older and that this jail time had a good influence. This combination of facts means they are less likely to commit another crime, as (B) claims.

A word of clarification might be in order. The two statements, A and B, each consist of a factual claim and a normative recommendation. At least (A) contains a normative recommendation if we drop "The authors concluded that" and state the subsequent phrase directly: "Older inmates should be held in prison even in the case of a prison population crisis." Here we do indeed have a contradiction, but I doubt anyone, including Nisbett, focuses on that directly. Instead, alert readers are likely to reject the normative advice given in (A) or given in (B), and look for the best policy in light of the facts stated in (A) and (B). This is what I have done above, rejecting the policy recommended in (A). The facts that lead to the pair of contradictory normative policies are themselves not in the least contradictory.

6   Taking Stock

It is time to take stock and draw some conclusions. I began by noting the existence of a great many answers to the Needham question, including Needham's own. But when we try to compare science in China with science in the West, we inevitably run into the problem of incommensurability. Of course, we might reject incommensurability as merely an alleged problem, not a genuine one. That might be the right response, but we need to provide some reason for doing so, since the case made by Kuhn and others is plausible.

One important thing we could say against Kuhn's claim is that Galileo's thought experiment, which is playing a central role here, is clearly intelligible to both the Aristotelian and the modern. Indeed, it is essential that this be so, since the final result, $H = L = H+L$, where H, L and H + L are the speeds of the heavy and light and combined bodies, follows from the earlier $H > L$ assumption, which is now seen to be a false claim. If Galileo's reasoning is correct, then the meaning of these terms cannot have changed in a significant way when we switch from Aristotelian to modern perspectives.

However, in addressing the Needham question, we do not have to concern ourselves with comparisons at the level of specific examples. We only need to answer the question, Why did the West scientifically race ahead of China? Detailed comparisons of specific theories is irrelevant.

When we are explaining the differences between Western and Chinese science, however, we do need to take more general methodological ideas into consideration. Thus, we must worry about such things as: Did both sides agree on the nature and importance of experimentation, logic, causation, and so on? Both groups might have agreed on the importance of such things, but then meant different things by them. Though Peng and Nisbett do not discuss their views in terms of incommensurability, they are in effect making a claim along this line. Basic reasoning, according to them, is very different among Westerners and Chinese; the former, to repeat, are analytic and reductionistic in their outlook, while the latter are dialectical and holistic. Anyone sympathetic to Kuhn would see that as a claim involving incommensurability.

In passing I should mention that others have been inspired by Nisbett's work. There is a recent trend in Western philosophy toward what is called experimental philosophy. Philosophers of language were persuaded several years ago to switch from an account of how names refer, which was initially proposed by Frege and Russell and which is known as the descriptions account of reference. They adopted a different account proposed by Kripke, known as the causal theory of reference. Kripke presented a series of imagined situations in which people found themselves persuaded that the causal theory of reference was correct. Machery et al. (2004), taking their cue from Nisbett, tested the intuitions of Americans and Chinese in a number of cases. They found that Americans favoured the causal theory, while Chinese favoured the descriptions account. They concluded that Kripke's argument for the causal theory was based on culturally conditioned intuitions and hence, drew the further conclusion that such intuitions are not to be trusted for that reason. I won't comment further on this example; I mention it only to show that Nisbett's influence is considerable, reaching out of psychology into both philosophy and the history of science.

7   The Needham Question

Peng and Nisbett claim that their account answers the Needham question. Unfortunately, they do not tell us precisely how, but reading between the lines we can see what they seem to be getting at, though it is not entirely coherent. In brief, modern science, which they take to be a good thing, was the product of analytic thinking. The analytic and reductionistic outlook of the West, they say, led to this success. This much seems clear in their account, even though it is only implicit. They are slightly incoherent, since they are at pains to praise the virtues of each cognitive style and even criticize Western thinking patterns from time to time. But the upshot of their view is that without analytic and reductionistic thinking, modern science would never have come into being. This is what gave the West the Scientific Revolution; it is what China lacked, depriving it of a similar revolution, in spite of being otherwise well prepared for it.

The Peng-Nisbett argument is not at all persuasive, because the West is not oblivious to context. We have seen this holism at work in a number of examples, from Aristotle, who held a kind of biological view of nature where parts are understood in terms of the role they play in the whole organism, to quantum mechanics, with its reliance on relational properties. The Scientific Revolution was a period when the West was particularly reductionistic and analytic in the sense of Peng and Nisbett, but even then people like Galileo were well aware of contextual factors and realized the need to address them, since his Aristotelian opponents took them seriously.

Finally, I think the biggest disappointment with Peng and Nisbett rests on their failure to distinguish scientists from non-scientists. Even if they are right about differences in Western and Chinese thinking generally, the inference from the general population to how their scientists think is a huge jump. Earlier I mentioned that American scientists are not religious, even though most Americans are. Their indifference to religion has a big impact on their thinking, especially in biology.

I am prepared to accept the Peng and Nisbett claim that Chinese and Western *scientific* thinking is different, but – and this is hugely important – the difference comes from within a paradigm; it is not transcendent. Let me stick with biology for a moment and with the idea of causation. This will provide a useful illustration. Aristotle has four kinds of cause and four corresponding types of explanation: material, formal, efficient, and teleological. The Scientific Revolution eliminated, at least in principle, all but efficient causation, typified by one billiard ball hitting another and causing it to move. Teleology lingered in biology, but only because no one knew how to eliminate it. Darwin's great achievement was in showing how the appearance of design and purpose could be achieved by so-called natural selection, which involves only efficient causation. Scientists in his day recognized the empirical virtues of Darwin's theory of evolution based on natural selection, but the great conceptual triumph was much deeper. He eliminated a type of causal mechanism, teleology, that was for many previous years something of an embarrassment to those who wanted to embrace a fully mechanical outlook.

A Peng-Nisbett type of study done in the mid-19th century would have found British biologists inclined to mechanistic thinking. They would have found the Chinese and the general British public more inclined to a teleological outlook, one that sees purpose and design in nature, where a holistic outlook is plausible. These are huge differences, but they are not longstanding cultural differences. They are internal to the general paradigm one is working within.

The upshot is obvious, but I will state it anyway. Peng and Nisbett have not made their case. The differences between the analytic Westerners and the dialectical Chinese is greatly exaggerated and, more importantly, in so far as the difference exists at all, it is tied to particular paradigms. It is not a stable feature of distinct cultures. We shall have to look elsewhere to finally answer the Needham question.

Bibliography

Brown, James Robert (2008). *Philosophy of Mathematics: A Contemporary Introduction to the World of Proofs and Pictures*. New York and London: Routledge.

Ferguson, Niall. "Too Much Hitler and the Henrys," *Financial Times*, April 10/11, 2010 (Life & Arts section).

Feyerabend, Paul (1975). *Against Method*. London: New Left Books.

Galileo Galilei [1638]. (*Dialogo*) *Dialogues Concerning Two New Science*, trans. by Crew and de Salvio (1914) London: MacMillan.

Kuhn, T.S. (1971). *The Structure of Scientific Revolutions*. Chicago: University of Chicago Press.

——— (1977). "Objectivity, Value Judgement, and Theory Choice," in *The Essential Tension*. Chicago: University of Chicago Press.

Lloyd, G.E.R. (2004). *Ancient Worlds, Modern Reflections: Philosophical Perspectives on Greek and Chinese Science and Culture*. Oxford: Oxford University Press.

Machery, Edouard, Ron Mallon, Shaun Nichols, Stephen P. Stich. "Semantics, cross-cultural style," *Cognition* (2004) 92(3): B1–B12.

Needham, Joseph (1954–). *Science and Civilization in China*, many volumes, Cambridge University Press.

——— (1981). *Science in Traditional China*. Cambridge, MA: Harvard University Press.

Nisbett, Richard (2003). *The Geography of Thought: How Asians and Westerners Think Differently . . . and Why*. Free Press: New York.

Peng, Kaiping and Richard Nisbett. "Culture, Dialectics, and Reasoning About Contradiction", *American Psychologist* (1999) 54(9): 741–754.

——— "Cultural and Systems of Thought: Holistic Versus Analytic Cognition," *Psychological Review* (2001) 108(2): 291–310.

Brilliant chapter

CHAPTER 4

Anthropocosmic Processes in the Anthropocene: Revisiting Quantum Mechanics vs. Chinese Cosmology Comparison

Geir Sigurðsson

1   Introduction

Western and Chinese philosophies may have started off with similar considerations, but later took quite different paths. Socrates and Plato were first and foremost concerned about the good life, both for an individual as such (ethics), and for a citizen as a part of a well-ordered whole (political philosophy). This concern led the more influential Plato into epistemological paths of thinking through which he sought to ascertain true knowledge of ethical matters, paths that were largely influenced by the pre-Socratic natural philosophers. Certainly, epistemological considerations feature prominently in Plato's works, but they are nevertheless merely an aspect of his rich and insightful philosophy of the good society and the good life. The same can be said about Aristotle, although his works certainly offer a wide range of scientific explorations. Thus, the ancient Greeks were primarily concerned about the nature of the good life in much the same way as the Chinese.

During the modern age, however, the age of Descartes and Kant, but, significantly, also the age of the scientific revolution, epistemological foci were given prominence as the proper subject matter of philosophy. Indeed, epistemology, the theory of how we know that we know what (we at least believe) we know, was now interpreted and understood as a methodological tool for science. Not only had the epistemological accent of modern Western philosophy been sealed, but it was taken for granted, e.g. by a disgruntled Martin Heidegger, that philosophy's theoretical, metaphysical and epistemological endeavour could be traced in a direct line back to the ancient Greek thinkers, which also significantly influenced the manner of studying their works.[1] As Heidegger put it,

---

1   As is well known, Heidegger held that philosophy had taken a "wrong turn" already with Plato and his insistence that truth is "subject to ideas, dependent on verifiability in a positivistic-scientistic fashion" (Ferkiss 1993: 165). He therefore ventured to redirect it by immersing himself in the study of the Presocratics and their quest for 'being'. Heidegger's

"the essence of modern science, which has become world-wide meanwhile as European science, is grounded in the thinking of the Greeks, which since Plato has been called philosophy." (Heidegger 1977: 157)

In China, on the other hand, epistemology never became a significant philosophical factor. To the (limited) extent that epistemological considerations were present, they tended to be secondary and often characterized by skepticism.[2] Scientific, or rather, technological innovations certainly took place in the Chinese Empire under a Confucian-led ideology, but they did so despite the lack of systematic epistemological theorizing.

A yet important aspect concerns the extent to which science enjoyed independent status in the West and in China. During the modern age in Europe, the natural sciences broke loose from both philosophy and religion, and became largely a separate sphere of human activity. While their formal justification was still Baconian in the sense that they were generally understood (and respected) as an effort to improve the human being's living conditions on the planet, their capabilities were not significantly limited by moral and political decisions. At least such demands did not become prevalent before the latter part of the twentieth century. In many ways, they were allowed to override the immediate and long-term moral interests of humankind. Morality had become secondary; knowledge and technological mastery primary.[3]

Philosophy in China developed in a period of almost three full centuries of incessant warfare and human misery, and it is therefore understandable that the early philosophical focus was on social order and ways to obtain social stability. When circumstances were different, however, i.e. in times of relative

---

disciple, Hans-Georg Gadamer, however, suggests that it is rather our understanding of the ancient masters that has been distorted by the scientist turn in modern philosophy. Hence, he offers a different interpretation that, at least in the field of humanities, eschews this epistemological insistence upon indisputable truths, observing, for instance, that "Aristotle contrasts 'ethos' with 'physis' as a field which, while not wholly disorderly, cannot be associated with the orderliness of nature, but the volatility and the limited regularity of human statutes and human conduct." (Gadamer 1990: 318)

2   This feature is particularly notable in Daoism. Both the *Daodejing* and the *Zhuangzi* are profoundly playful with language as a human construction, twisting and turning common references of words and expressed valuations. A good case in point is chapter 2 on language and reality in the *Zhuangzi*. However, Confucius's notion of *zhengming*, or 'using words appropriately', also implies considerable wariness of the potential deception involved in language, albeit predominantly from an ethical or prescriptive point of view.

3   Note, for instance, how Immanuel Kant's *Critique of Pure Reason* from 1781 is normally studied, even still today. Its second part, dealing with the *meaning* of his epistemological considerations for human life, tends to be largely neglected.

peace, it should have been a stimulant for other foci. This was certainly the case with Daoism and Confucianism during the Han, Tang and Song dynasties, perhaps even during the first decades of the Ming. But during the Ming, when economic and social circumstances underwent enormous changes that would have required appropriate responses from both intellectuals and political leaders, Confucianism failed to produce these responses. One reason is of course the long-standing Confucian lack of interest in, even contempt for, commercial affairs and economic profit. But the divide between, on the one hand, an idealized form of government and organization and a fast changing reality, on the other, further contributed to China's stagnation during and after the Ming dynasty. Helplessly facing an administration largely in the hands of corrupt eunuchs of the inner court who despised the educated class, the Confucians at the end of the Ming turned their attention away from the present and future evolution of society, and inward into the past, towards a pedantic, dogmatic and reactionary view of ritual and correct behaviour.

During the Qing, Confucian scholars found themselves in an even more complicated dilemma. They had, just like the Qing emperors, repudiated the idealist philosophy initiated by Wang Yangming for stimulating the selfishness and moral corruption that brought down the dynasty. (Spence 2013: 100f.) However, they were also incapable of sharing the foreign Manchu rulers' adoration of Song neo-Confucian orthodoxy. And lastly, the Manchu emperors exerted rigorous control over scholarship in order to avoid the publication of anti-foreign writings as well as potentially revolutionary activities. Not many options seemed available. The way most scholars found a way out of this dilemma led them in fact further back, all the way to the original Confucianism of the Zhou dynasty through Han dynasty sources, whereby they also introduced a rigorous methodology of textual criticism, the so-called "evidential research" (kaozheng xue). Unfortunately, this revival of the antiquity did not produce a revival of Chinese culture comparable to the revival enjoyed in the West following the rediscovery of classical texts during the European Renaissance. "Evidential research" involved a disapproval of speculation and demand for "hard facts", which may sound as a form of scientific empiricism, but which gradually narrowed itself down to a rigorous and rather obscure textual analysis, such that many a group of scholars was "... so rigid in its view of the ancient commentaries of the Eastern Han as to preach that 'the ancient teachings cannot be revised' and one can only 'maintain conformity to the family statutes of the Hans.'" (Zhu 1990: 126) Seeking their own identity in the classical sources, the tendency of Ming-Qing Confucianism was towards a further reification of the Confucian practices, including, of course, education and its "ingrained" innovative force. Needless to say, the education system suffered in a comparable manner. It is therefore fair to say that from the Qing dynasty onwards, long before the civil service examinations were abolished and Confucianism officially denounced in the twentieth century, Confucianism ceased to be a creative catalyst in Chinese educational and scientific practices.[4]

Thus during the Ming and Qing dynasties in China scientific activity was subordinate to moral and political considerations as determined by Confucian scholars and their superiors, paramount of whom was the emperor himself, which is significant during the long reigns of Kangxi and Qianlong. The odd and contradictory situation for scholars during the latter half of the Qing dynasty, being overseen by emperors who were just as eager to promote Confucian scholarship as they were reactionary in their interpretation of it, was unlikely to produce bold experimentation that could have led to some kind of scientific revolution.

These constitute the main reasons why systematic Chinese science on par with Western science was never born. The explanations provided may be some-what simplified, but the point here is that there are both internal and external reasons for the differences between science/philosophy in the West and China. The internal reasons concern the specific nature of Chinese thought, while the external reasons are socially and politically determined. Overall, it is fair to say that a clear distinction between fact and value, which began making its mark in the West during the modern age, never really emerged in China. This means that social, political, and, perhaps more significantly, philosophical or ethical considerations were always in control over scientific activities and development.

While this is not the core of my topic in this essay, it nevertheless has a bearing on it. My aim in these pages is to embark upon an investigation of the contemporary relevance of classical Chinese thinking by considering Chinese philosophy or the classical Chinese worldview and philosophical or cosmo-logical implications of relatively recent scientific discoveries. Certainly, tradi-tional Chinese and Asian, notably Buddhist, worldviews have been compared with such implications before.[5] This discourse has been located outside the mainstream, and as far as I understand, it has also been disputed by many (mainstream) scholars. But I believe that a confrontation with the Daoist/ Neo-Confucian worldview in this respect may possibly lead us to some produc-tive and interesting implications regarding the status and the conditions of human living in the world. Something can, I think, be learned from the Chinese

---

4   A further discussion of this issue is found in Sigurðsson (2010), pp. 69ff.

5   The most notable are without doubt Fritjof Capra's *Tao of Physics* (1975) and Gary Zukav's *The Dancing Wu Li Masters* (1979).

perspective, something that actually relates both to our current circumstances as living, thinking and acting creatures on planet Earth and the new cosmology that has been emerging on the basis of relatively recent discoveries in physics.

In the following, I will begin by discussing some of the more intriguing cosmological implications that have been drawn from quantum physics. Then I shall provide an admittedly criminally short account of Chinese cosmology as it unfolded from ancient times to the Neo-Confucianism dominant in China from approximately the 12th to the 19th century. While these two have previously been compared by a number of intellectuals, my intention is to work towards a path that takes us beyond mere comparisons. With this in mind, I will draw briefly upon a related example in the fourth and final section that may indicate such a path. It concerns a call made by a number of scientists for a new way of approaching environmental issues in light of current affairs. The expression of this new 'consciousness', I argue, is in accordance with the philosophical, including even *moral*, implications that we *ought to* draw from quantum theory, and which could be further enriched by Chinese cosmological insights.

2   Cosmological Implications of Quantum Mechanics

Quantum mechanics has important philosophical (cosmological, metaphysical or ontological) implications about the nature of reality. But while the forerunners of quantum mechanics recognized this themselves, physicists have been reluctant to work out these implications in detail, possibly because they have been too preoccupied with its practical applications. Quantum mechanics has been described by the Nobel laureate physicist Murray Gell-Mann as "that mysterious, confusing discipline which none of us really understands but which we know how to use." (Kumar 2009: xvii) Certainly, without quantum mechanics, we would not have computers or washing machines, mobile phones, microwave ovens or, for that matter, nuclear weapons. But for some the philosophical inferences to be drawn from quantum mechanics may also be too radical to be acceptable. They most certainly were for Albert Einstein, who was never willing to look them in the eye, and rejected them with his well-known statement "God does not play dice with the universe". But if such an entity exists, then He or She or It indeed seems to play dice, for the philosophical implications of quantum mechanics would seem to require no less than a 'shift of paradigms', to use a notion coined by Thomas Kuhn, involving a departure from Newton's (and still our) clockwork cosmos of strict causal relations, determinism, locality and objects independent of perception.

While Niels Bohr was the one who persistently pursued the philosophical meaning of quantum theory, and famously declared that "anyone who is not shocked by quantum theory has not understood it" (Barad 2007: 254), Werner Heisenberg was probably the first to express himself clearly on the issue. He said: "But what is wrong in the sharp formulation of the law of causality, 'When we know the present precisely, we can predict the future,' is not the conclusion but the assumption. Even in principle we cannot know the present in all detail." (Kumar 2009: 248f)

A breakthrough in the development of quantum mechanics was Heisenberg's uncertainty principle (or "indeterminacy principle", as he sometimes called it), which states that a precise determination of both the position and the momentum of a particle is impossible. The more precisely one can be measured, the less precisely the other can be predicted or known. The reason for this is the wave-particle duality, first discovered to apply to light, but later to apply to electrons as well, and therefore to that which we commonly refer to as 'matter'. Depending on how it is observed, it will reveal itself as either wave or particle but never as both at the same time. Reality, in this sense, is both energy and solid matter. This indeterminacy of reality is embodied by 'superpositions', i.e. that a given aspect of reality exists in all the theoretically possible configurations of its properties simultaneously, but when it is measured or observed, it reveals only one of its possible configurations.

A conceptual framework for describing and accommodating this paradoxical nature of wave-particle duality was Bohr's 'complementarity'. "The wave and particle properties of electrons and photons, matter and radiation, were mutually exclusive yet complementary aspects of the same phenomenon. Waves and particles were two sides of the same coin." As Bohr himself argued, "evidence obtained under different conditions cannot be comprehended within a single picture, but must be regarded as *complementary* in the sense that only the totality of the phenomena exhausts the possible interpretation about the objects." (Kumar 2009: 242)

Any interaction between an observer (including of course the equipment used) and microphysical objects require the exchange of at least one quantum of energy. This led Bohr to affirm the "impossibility of any sharp distinction between the behaviour of atomic objects and the interactions with the measuring instruments which serve to define the conditions under which the phenomena appear." (Kumar 2009: 244) This means that the traditional separation of observer and observed in classical physics no longer applies.

An implication of this dual nature of reality is that whether we choose to understand the smallest aspects of reality as particles or waves, the law of causality, as we find in Newton's theory, is not an accurate description of reality, at least not in the microscopic dimension of the world. This further means that a deterministic account of the universe is impossible, as we will never be able to know all the factors at work in any circumstance, and can therefore never predict what will happen with 100% accuracy. We can only speak of tendencies and probabilities. What is striking about this account is that this 'shortcoming' has nothing to do with less than adequate means of measurement, but is due to the nature of being (or becoming). Any improvement in the equipment used to measure reality will not change the fact that world-operations cannot be determined – it is an intrinsic feature of reality.

Further, the 'objects' under investigation are not independent entities. "There is no quantum world," as Bohr said, "there is only an abstract quantum mechanical description." (Kumar 2009: 251) Quantum mechanics is not in this sense a 'model' of the world, as classical physics purported to be, but a tool for accessing it from certain (and therefore by no means 'objective' and 'value-free') perspectives.

After this all-too short and inadequate description of the most dramatic ontological results of quantum mechanics, I will now move to an equally incomplete summary of its further implications based on physicist and philosopher Karen Barad's recent and most fascinating study of "quantum physics and the entanglement of matter and meaning" as implied in the subtitle of her book *Meeting the Universe Halfway*.[6]

In short, Barad argues that Bohr's theories contradict some fundamental modernist assumptions about our relations with reality: first of all, representationalism, i.e. the idea that words and objects, or meaning and matter, have clearly separable spheres of being and that 'objective' reality can be represented with words without difficulties. There are no individually determinate entities to be discovered. All we have, Barad argues, are phenomena arising from the *intra-action* (as distinct from 'interaction') of 'objects' and 'measuring agencies', and these phenomena are the only available conceptual schemes involving determinate boundaries and properties. (Barad 2007: 127f) Meaning arises when specific 'agential intra-actions' take place, determining the boundaries and properties of the 'components' of phenomena. In this case, "particular material articulations of the world become meaningful." (Barad 2007: 333) In other words, meaning arises necessarily as a co-creation between human and world, but there is no objective reality to be 'discovered' as such.

---

6   I am also indebted to my friend and colleague Björn Thorsteinsson for sharing with me his illuminating paper on this topic, "On the Ontology of Quantum Mechanics", which appeared in Icelandic in 2010.

Secondly, quantum mechanics contradicts metaphysical individualism or individual atomism, i.e. the belief that the world is composed of individual and clearly separated units containing certain inner properties or substances that have nonrelational properties. As the physicist N. David Mermin has said: "Correlations have physical reality; that which they correlate does not." (Barad 2007: 332) A relational ontology arises from Bohr's ideas. Barad suggests that 'phenomena' should be taken as the primary unit, but "phenomena are the ontological inseparability of intra-acting 'agencies' ... they are the basic units of existence." (Barad 2007: 333) This also entails the notion of 'quantum entanglement', giving rise to a profoundly complex relationality between the various phenomena in the world.

Thirdly, and following from the inferences above, quantum mechanics reveals that a distinct separation between 'observer' and 'observed', between 'subject' and 'object', is not possible. This leads Barad to work towards a 'posthumanist' approach according to which the human perspective is merely one out of infinite possible interpretations of reality, and not at all a non-natural process. On the contrary, while humanism focuses on the human as something exceptional, Barad's "posthumanist elaboration of Bohr's account understands the human not as a supplemental system around which the theory revolves but as a natural phenomenon that needs to be accounted for within the terms of this relational ontology." (Barad 2007: 352) In other words, this vision requires a breakdown of the classic nature-human, or nature-nurture, dualist dichotomy by understanding human beings and their actions as perfectly natural phenomena.

The discussion provided above is of course a very rudimentary account of a most complex theory leading to a truly revolutionary understanding of the reality of which we are all a part. "It is all quite mysterious", as Richard Feynman stated, "and the more you look at it the more mysterious it seems." (Barad 2007: 254) But while the theory is most certainly complex – not least to a non-expert such as myself – the world it describes may not necessarily be so mysterious as Feynman states – at least not to Chinese cosmologists to whom we shall now turn.

3   Chinese Cosmological Visions

It would be a daunting task, to say the least, to intend to provide anything resembling a comprehensive account of classical Chinese cosmology in this discussion. Therefore, I shall limit myself to the attempt to identify some of its consistent characteristic features that are relevant to the issue at hand. For the sake of simplification, I will allow myself to blend, in my discussion, Daoist and Neo-Confucian cosmological views. Certainly, there are differences between the two, but they significantly overlap, both because they draw mostly from the same ancient sources, but also and no less because the Neo-Confucian cosmological vision that was formed in the 10th to 13th centuries was heavily influenced by Daoist cosmology.[7]

To begin with, it is of significance that the earliest Chinese cosmological views were based on an ancient manual of divination, the *Classic of Changes* or the *Yijing*. It is of significance, because the starting point was precisely a dynamic 'system', for lack of a better word, of symbols used to establish some kind of order when dealing with the uncertainties of everyday human life. This is not to say that the cosmological vision that was produced did not take the structure of nature and reality into account – otherwise, it would most likely have been disproven before long as useless. But the perspective adopted to approach cosmological considerations was not detached but 'interested', one that was concerned with its meanings and implications for the human condition. In this sense, it was, to use a term coined by Tu Weiming, 'anthropocosmic'. Despite the many changes or additions made to the system in time (and depending on the various schools and thinkers dealing with it), this essential feature or dimension of the system was never lost. In other words, a drive towards a kind of 'objectivity' (as opposed to a 'subjective' dimension) was absent.

The absence of a subject-object dichotomy is a feature of a more far-reaching absence of a strict kind of dualism containing mutually incommensurable items, such as matter-spirit, body-soul, animate-inanimate, appearance-reality, and so on. Certainly, a kind of dualism is presented and utilized as a symbolic illustration and identification of continuously changing properties in a transforming world. But it is a yin-yang informed dualism according to which the continuous aim of the ever-flowing process in question is the establishment of a harmonious relation between complementary characteristics expressed through conceptual contrasts.

A good and important example of this is *qi*, the basic substance, entity, stuff of reality. The lack of appropriate vocabulary for *qi* within Western linguistic and philosophical frameworks has had the consequence that *qi* is described in a wide variety of ways, as 'psychophysical stuff', as 'matter-energy', as 'breath', and even using ancient Greek philosophical vocabulary as *pneuma*. *Qi* is a

---

7   And of course to Buddhism as well, although the Neo-Confucian thinkers were more reluctant to draw from Buddhism as they sought to curb its (moral) influence in Chinese society.

pervasive notion in the classical Chinese vocabulary, found in Daoist, classical Confucian and Neo-Confucian writings, perhaps precisely reaching its heights in the Neo-Confucian philosophy as the second aspect of reality along with *li*. By itself, *qi* denotes a dual feature of being; both that which we identify as 'matter' and that which we understand as 'energy', since the continuous changes at work in reality are 'immanent', i.e. occurring by themselves as expressed by the notion of *ziran*, or 'that which is so by itself', literally 'self-so-ing'. Matter, in this sense, contains an inner drive, is a dynamic autopoietic phenomenon, which also implies that a clear metaphysical distinction between animate and inanimate beings cannot be sustained. Life is therefore contained in 'matter' as one kind of dynamism characterizing the ever-changing reality. *Qi* is both source of life and a pervasive feature of being, or, more appropriately, of becoming. What ensues from this, among other things, is a breakdown of a clear distinction (or dualism) between human beings and their environment, between culture and nature, and, from a scientific point of view, between subject and object.[8] This is not to say that making such distinctions is meaningless. On the contrary, they can be useful for heuristic purposes. But within a yin-yang ordered world in flux, they will be treated as contingent, temporary, reversible and unreifiable.

The anthropocosmic orientation in the Neo-Confucian philosophy compelled its thinkers to go even further than that. Presumably, they felt that they needed something more than *qi* to account for some sort of regularity, or, indeed, *meaning* in the world. *Qi* was therefore combined with *li*, a notion of 'pattern' or 'patterning', to indicate a meaningful dynamic structure of reality. How does *li* make reality meaningful?

Li is that which provides every single thing, composed of *qi*, with an identity. It is the pattern within things that distinguishes them from other things and thus differentiates them. *Li* is consistently formulated within a yin-yang kind of sensibility as something that cannot exist without *qi*, just as *qi* is incapable of existing independently of *li*. (Chen 1999: 21) In this sense, it would seem that there are innumerable kinds of *li* in the world, and this seems to be confirmed by the Cheng brothers who stated that "all things are endowed with *li*." (Hou et al. 2005: 143) However, there is also a *li* of the totality of things, *tianli*, a pattern or structure of the world as a whole. The *tianli* is to some degree the Neo-Confucian substitute for the Daoist *dao*, the world process in its entirety, while it seems to refer more specifically to the dynamic pattern within the process.

Now *tianli* (and, for that matter, *dao*) are not merely descriptive notions. They are also *pre*scriptive in the sense that a certain emulation of the natural processes ought to inform us of the proper way to lead our lives. In Daoism,

---

8 Cf. e.g. Tang (1988).

emulating natural processes (as expressed through *dao*) seems to be regarded predominantly as expedient and wise. Someone who fails to live according to the natural process of *dao* will simply lose out (in a variety of senses). In Neo-Confucianism, however, it goes beyond mere expediency by presuming that reality is inherently benevolent or good.[9] *Tianli* is thus a 'good' pattern that we should study and emulate in order to improve ourselves. Therefore, Zhu Xi (1130–1200) interpreted the ancient manual of personal cultivation, the *Great Learning* (*Daxue*), in such a way that the starting point of cultivation consists in extending one's knowledge by investigating things (*gewu zhizhi*). By studying things, one developes a sensibility for the *li* inherent in things, cumulating in a sensibility of *tianli*.

How should we describe such sensibility? In the Confucian vocabulary, and on a human level, it means that we develop our inborn natural tendencies, our *xing*, towards becoming a consummate person, or *ren*. And in fact, *li* of the human being is precisely identified with *xing*, the inborn natural tendencies. *Xing* is the human *li*, the particular 'pattern' that provides human beings with their 'humanness'. When *xing* is successfully transformed to enter the path towards *ren*, one's sensibility could be said to approach *tianli*. Remember that personal cultivation (towards *ren*) begins with the study of the *li* of things in order, at some point, to realize the *li* of the totality, the *tianli*. In this sense, *tianli* and *ren* are intimately related as macroscopic and microscopic expressions of the totality of being and the human being, that which brings nature and the human into unity.

All taken together, the Daoist/Neo-Confucian cosmological vision seems to present us with the following characteristics:

First of all, reality is a ceaseless process of change, a "continuous cosmogony." (Linck 2001: 14.) This has a number of consequences for our approach to reality. Unchangeable principles, laws or anything resembling Platonic forms are *a priori* excluded as a logical impossibility. The only thing that does not change in this process is change itself. While some level of regularity and orderliness can be discerned in the process, nothing can be absolutely 100% certain. This world cannot be conceived as deterministic, not even as one containing

---

9 The usual Chinese rendering of 'goodness' would be *shan*, which, however, has also been interpreted as 'efficacy' or 'deftness' to underline the contextuality of 'goodness,' that 'good' is always 'good for' or 'good to' something. There is no abstract notion, no Platonic form of '*the Good*,' in classical Chinese philosophy (cf. Hall and Ames (1998), p. 278). Hence the difference between the Daoist and the Neo-Confucian outlooks may not be as clear as it may seem at first sight.

necessary natural laws, but will at most be expressed through tendencies with various degrees of likelihood.

Secondly, everything in this world is relational and all things interpenetrate one another. Why is this the case? Because there is no unchangeable essence constituting things, they are only what they are by virtue of their relations with other things. Epistemologically, this means that they can only be approached and known from perspectives. There *is* only perspectival knowledge and understanding. This entails that *any* understanding of or insight into another thing/phenomenon is a result of the unique relations between knower and known. Ontologically, however, this means that everything affects everything else. (Linck 2001: 20) The world is a web of relations whose understanding excludes reductionism. One change will affect all the others, if in most cases only minimally, but nothing is perfectly isolated.

Thirdly, and consequently, the natural and the human are intimately related as a complementary and, in a certain sense, *moral* unity in an anthropocosmic scheme that sees the human dimension as an integral part of the cosmic processes. The traditional portrayal of this union in the Confucian tradition is the triangular relationship between 'heaven' (*tian*), 'earth' (*di*) and 'human' (*ren*), indicating some sort of mutual protection or responsibility that secures the preservation of all. The notion of 'intimacy' implies a level of emotional attachment modeled on familial relations but extended to the natural realm, other living beings, and even phenomena traditionally held to be inanimate. Such a wide extension signifies the culmination in the personal cultivation of the consummate person (ren). In the *Western Inscription* (*Ximing*), the Neo-Confucian thinker Zhang Zai (1020–1077) formulated an expression of all this that could be taken as a kind of Neo-Confucian manifesto:

> Yang is the father; yin is the mother. And I, this tiny thing, dwell enfolded in Them. Hence, what fills Heaven and Earth is my body, and what rules Heaven and Earth is my nature. The people are my siblings, and all living things are my companions. My Ruler is the eldest son of my parents, and his ministers are his retainers. To respect those great in years is the way to "treat the elderly as elderly should be treated." To be kind to the orphaned and the weak is the way to "treat the young as young should be treated." The sage harmonizes with Their Virtue; the worthy receive what is most excellent from Them. All under Heaven who are tired, crippled, exhausted, sick, brotherless, childless, widows or widowers – all are my siblings who are helpless and have no one else to appeal to. To care for them at such times is the practice of a good son. To be delighted and without care, because trusting Them, is the purest filial piety. . . . Riches, honor, good

fortune, and abundance shall enrich my life. Poverty, humble station, care, and sorrow shall discipline me to fulfillment. Living, I compliantly serve Them; dead, I shall be at peace.[10]

4   Entering the Anthropocene Era: Can Chinese Philosophy Make a Difference?

The usual, and probably justified, question when reaching the end of this sort of comparative investigation is 'so what?' Several authors, e.g. Fritjof Capra and Gary Zukav, have already pointed out the similarities between the worldviews found in quantum mechanics, on the one hand, and classical Chinese philosophy, on the other, though they, as well as many others, have found more similarities with classical Buddhist metaphysics.[11] As intriguing as such comparisons may be, what we gain from them is perhaps not as obvious. What is at least clear, however, is that while there may be significant similarities between these accounts, neither the Daoist/Confucian nor the Buddhist thinkers had arrived at these results by sophisticated scientific experimental means. East Asians did not discover quantum mechanics. During the colonial era, Newtonian cosmology was imported from the West into these cultural areas, and for the most part, replaced older cosmologies and worldviews.

So why should their worldview be relevant? I believe that what we can learn from the Chinese (and Buddhist) vision is a perspective and possibly also vocabulary that may contribute to our endeavour to reconsider and even reconceptualize the new circumstances in which we currently find ourselves. Quantum mechanics has revealed to us that the classic Western dualist worldview and the vocabulary entailed by it is inadequate as a formulation of the world "as it really is". Dualist concepts such as matter-spirit, subject-object and cause-consequence do not realistically mirror reality. The relational ontology deriving from quantum mechanics is further an important corrective to the Newtonian world of determinate borders, properties and strict causality. Not only have these modernist views been shown to be inaccurate representations of reality, but they may even turn out to hamper ways for human beings to organize themselves appropriately within their environment, and therefore prove to be downright pernicious for the future evolution of being-in-the-world, whether human or other beings.

---

10   Zhang (2006). The quoted phrases are from the Confucian 3rd century BCE work *Mengzi*.
11   Capra's *Tao of Physics* (1970) and Zukav's *The Dancing Wu Li Masters* (1979) are both groundbreaking but certainly controversial works.

This accords with the perception of a number of European scientists who are calling for new approaches to environmental issues in the light of the emerging and formidable impact of human activity on planet earth: "It has created a completely novel situation that poses fundamentally new research questions and requires new ways of thinking and acting." (Palsson et al. 2013: 2) This emerging epoch, in which human activity must "be considered a 'driver' of global environmental change", has been referred to as the 'Anthropocene'.[12] Asking for a healthy integration of both natural and human sciences in environmental studies, these scientists argue that the human being cannot any more be considered apart from nature, and "the environment must be understood as a social category". (Palsson et al. 2013: 4) This would seem to require quite novel perspectives on the nature-human relationship and may even challenge us to think about the categories of nature and human in new terms:

> Nature has often been presented as one half of a pair – nature/culture, natural/social, and so on. This is still echoed in some earth-system notions that are fundamentally dualistic, 'linking,' 'connecting,' and 'coupling' the two systems of the earth and humans as if they were different realities. But recently, environmental discourse has increasingly emphasized the need to move beyond the stark dualism of the natural and the social. (Palsson et al. 2013: 7)

Some of the main reasons for such an emphasis are the outcomes of empirical research, suggesting that the human impact on natural occurrences and even genetic conditions of both humans and animals is considerably more than hitherto believed. Thus, a strict demarcation of the human vs. the natural is increasingly seen by scientists as an unrealistic reflection of the real state of affairs.

Importantly, this awareness of the new 'human condition', to quote Hannah Arendt, has profound ethical implications. A classic modernist approach to the environment purely as a resource for human consumption is no longer viable. As Palsson et al. point out: "We are only part of a complex network of elements

---

12   Palsson et al. (2013), p. 2. Cf. also Ellis and Haff (2009), p. 473: "We live in the Anthropocene: For better or for worse, the Earth system now functions in ways unpredictable without understanding how human systems function and how they interact with and control Earth system processes. Regardless of whether this transition from the Holocene (generally thought of as the past 12,000 years) to the new epoch of the Anthropocene will ultimately be for the better or for the worse, the Earth system will not be returning to a preanthropogenic state for the foreseeable future."

and relations that make up planet earth, but we are the only part that can be held responsible." (Palsson et al. 2013: 9) They go on to refer to feminist theory and ethics of care as potential alleviators of this rigid modernist approach, which, despite an awareness of the need for change, retains us in an economic model whose aims are directly antagonistic to the environmental situation. But it is well worth investigating whether the Asian cultural and philosophical sensibilities, having operated for a long time in a much more 'responsive' conceptual relationship with nature, may have something to teach us. The yin-yang kind of dualism seems for instance much more realistic than our classical Platonic-Christian-Cartesian dualism, and it would seem to stimulate a 'softer' and certainly more moral relationship between human and world, thus being more likely to contribute to the formation of a culture of sustainability.

True to the commitment to value neutrality of Western science, quantum mechanics refrains from the attempt to make ethical inferences. The question is, however, whether there are not some such inferences that logically derive from its descriptive framework, that, in other words, *ought to be made* on its basis? This, I believe, is Barad's argument. Quantum mechanics reveals to us a world that undermines the modernist notion of an atomic individual, shows that our actions are always of consequence for the whole, and should therefore lead to a stronger sense of responsibility, an "ethics of mattering," as she refers to it. (Barad 2007: 391ff)

We appear to be entering a new era: an era in which we are compelled to acknowledge that our mere existence in this world imposes on us a responsibility for our actions and decisions, as these can never be seen as merely self-regarding. While the worldviews produced by quantum physics and the Chinese philosophical perspectives are certainly not identical, they nevertheless share a number of factors that would seem to lead us to draw important ethical inferences about human action in the world. The close and logical relationship between a scientific view of the world and a desirable moral life was never disputed in China. In the modern West, this is quite new (while certainly present in ancient Western philosophy), and we must find ways to deal with it. But do we need Asian philosophy for this task? Do we not have sufficient resources in Western philosophy to conceptualize these new relations and new responsibility to the world, perhaps in Heidegger's philosophy, phenomenology and/or post-structural/post-colonial thought? To a large extent I think the answer is 'yes.' All these schools of thought, if we can refer to them as such, have attempted to find ways to formulate reality by taking into account the human being's dynamic intra-active relationship with it, also as an observer who affects reality by the simple act of observing. However, not only are Asian philosophical perspectives helpful to open up these vistas; they have also exerted considerable influence on the Western philosophical approaches in question. In an increasingly globalized world, it would simply be unwise not to seek to profit from the Asian philosophical experience. Rather than continuing the modernist project based on "the dualism of nature and society, the notion of objective science, and the assumption of linear control," (Palsson 2006: 72) an anthropocosmic approach to the anthropocene, inspired by Daoist/Neo-Confucian philosophy, can no doubt contribute to a healthier and more sustainable culture for the future of life on earth.

Bibliography

Barad, Karen (2007). *Meeting the Universe Halfway: Quantum Physics and the Entanglement of Matter and Meaning.* Durham and London: Duke University Press.

Capra, Fritjof (1975). *The Tao of Physics.* Boulder, Co.: Shambhala.

Chen, Lai. "The Concepts of Dao and Li in Song-Ming Neo-Confucian Philosophy," *Contemporary Chinese Thought* (1999) 30(4): 9–24.

Ellis, Erle C. and Peter K. Haff. "Earth Science in the Anthropocene: New Epoch, New Paradigm, New Responsibilities," *Eos* (2009) 90(49): 473–474.

Ferkiss, Victor (1993). *Nature, Technology & Society. Cultural Roots of the Current Environmental Crisis.* New York: New York University Press.

Gadamer, Hans-Georg (1990). *Wahrheit und Methode. Grundzüge einer philosophischen Hermeneutik* (6th ed.) Tübingen: J.C.B. Mohr (Paul Siebeck).

Hall, David L. and Roger T. Ames (1998). *Thinking from the Han: Self, Truth, and Transcendence in Chinese and Western Culture.* Albany: State University of New York Press.

Heidegger, Martin (1977). "Science and Reflection," in *The Question Concerning Technology and Other Essays.* Translated with Introduction by William Lovitt. New York: Harper and Row, pp. 155–182.

Hou, Wailu et al. (2005). *Song Ming lixue shi* [*History of Li-learning in the Song and Ming Dynasties*], vol. 1. Beijing: Renmin chubanshe.

Kumar, Manjit (2009). *Quantum: Einstein, Bohr and the Great Debate about the Nature of Reality.* London: Icon Books.

Linck, Gudula (2001). *Yin und Yang: Die Suche nach Ganzheit im chinesischen Denken.* Munich: Verlag C.H. Beck.

Palsson, Gisli (2006). "Nature and Society in the Age of Postmodernity," in A. Biersack and J. Greenberg (eds.), *Reimagining Political Ideology.* Durham: Duke University Press, pp. 70–93.

Palsson, Gisli, et al. "Reconceptualizing the 'Anthropos' in the Anthropocene: Integrating the social sciences and humanities in global environmental change research." *Environmental Science & Policy* (2013) 28: 3–13. http://dx.doi.org/10.1016/j.envsci.2012.11.004

Sigurðsson, Geir (2010). "Towards a Creative China: Education in China," in Fan Hong and Jörn-Carsten Gottwald (eds.), *The Irish Asia Strategy and Its China Relations.* Amsterdam: Rozenberg Publishers, pp. 61–77.

Spence, Jonathan (2013). *The Search for Modern China* (3rd edition). New York and London: W.W. Norton & Company.

Tang Junyi (1988 [1937]). "Zhongguo zhexue zhong ziran yuzhou guan zhi tezhi" [The Special Characteristics of Natural Cosmology in Chinese Philosophy], in *Zhongxi zhexue sixiang zhi bijiaolun wenji* [*Collected Essays on Comparative Chinese-Western Philosophy*], Taipei: Xuesheng shuju.

Thorsteinsson, Björn (2010). "Verulegar flækjur: um verufræði skammtafræðinnar." [On the Ontology of Quantum Mechanics]. *Vísindavefur: ritgerðarsafn til heiðurs Þorsteini Vilhjálmssyni sjötugum, 27. september 2010.* Reykjavík: Hið íslenska bókmenntafélag, pp. 1–9.

Zhang, Zai (2006 [11th c.]). *The Western Inscription.* Translated by B.W. van Norden. http://faculty.vassar.edu/brvannor/Phil210/Translations/Western%20Inscription.pdf

Zhu, Weizheng (1990). *Coming Out of the Middle Ages.* Translated by Ruth Hayhoe. Armonk and London: M.E. Sharpe, Inc.

Zukav, Gary (1979). *The Dancing Wu Li Masters: An Overview of the New Physics.* New York: William Morrow and Company.

CHAPTER 5

Ibn al-Haytham and the Experimental Method

Cecilia Wee

1 Ibn al-Haytham's Contribution to the Experiment

Any assessment of contributions to the development of science in the West must surely include consideration of the scientific work of the great Alhazen, or Ibn al-Haytham. Ibn al-Haytham is perhaps best known for his groundbreaking work on optics. His *Kitab al-Manazir*, or *Book of Optics*, profoundly shaped the subsequent trajectory of optics as a discipline. This work is known to have influenced, through its Latin translation, the works of Roger Bacon and Kepler and set the foundation for Western work on optics for the next few centuries.[1] But Ibn al-Haytham's achievements were not confined to optics. He was indeed a polymath, who also made significant contributions to other fields such as mathematics and geometry, engineering, psychology, anatomy, and physics.

This study will, however, not specifically examine the nature of his contributions in any of these varied areas. I would instead like to examine another aspect of his contribution to Western science – namely, the *means and methods* that he employed in order to effect his significant contributions in these various fields. In his study blending history and philosophy Ibn al-Haytham has been called, with good reason, the 'first scientist' by author Bradley Steffens (2007). A trawl through Ibn al-Haytham's scientific work reveals amply the emphasis he placed on careful and detailed empirical work. Indeed, one particular contribution to scientific method that is commonly associated with Ibn al-Haytham is his role in developing the method of the experiment. It is this specific contribution by Ibn al-Haytham's that I will now explore.

Earlier writers have lauded his contribution to this facet of science. For example, George Sarton, the father of the history of science, notes that he was 'the best embodiment of the experimental spirit in the Middle Ages'. (Sarton 1927: 694) Sarton's comment is a compliment of no mean order, and this study will try to characterize the precise nature of Ibn al-Haytham's contributions to the development of experimental method.

---

1 See Crone (1999), p. 30.

However, Ibn al-Haytham's contribution to the experimental method is arguably a significant contribution to modern science *only if* the experimental method itself is in fact an important component of modern scientific procedure. Recently, there has been considerable argument aimed at showing that experiment does not play the important role in science that had previously been ascribed to it. In particular, it is argued that experiment does not, as had previously been thought, play any crucial role in the confirmation and disconfirmation of scientific theories. For example, philosophers such as Kuhn and Feyerabend have argued that there is no such thing as theory-independent observation, and that what one 'sees' in an experimental set-up may well be shaped and determined by one's theoretical beliefs. Hence, an experiment cannot be used to *independently* confirm or disconfirm a scientific theory. (Kuhn 1970; Feyerabend 1958)

In view of such considerations, a re-evaluation of Ibn al-Haytham's contributions to the overall scientific endeavor would be quite timely. This essay offers such a re-evaluation, and will have the following structure. Section 1 will focus on Ibn al-Haytham's own experimental work. I begin by examining the characteristics of observations which are made under experimental conditions, contrasting them with those that come in non-experimental situations. I then use this contrast to evaluate the *specific* nature of Ibn al-Haytham's contribution towards the experiment. Section 2 then examines the claim that experiment itself plays no significant role in scientific research and practice. Here, I shall argue that, while the role played by experiment in science may not always be as straightforward as previously portrayed by inductivists, positivists, and logical empiricists, experiment can still play a significant role within scientific endeavour. Indeed, as I shall show, Ibn al-Haytham's own work exemplifies that this is possible. Thus, his contribution to the experimental approach has to be deemed significant for overall scientific method. Section 3 provides a brief summary and conclusion to this study.

2   Experimental vs. Non-Experimental Observations

In early 20th century accounts in the philosophy of science, a clear-cut distinction was often drawn between scientific theories/hypotheses and scientific observations. The nature of the support provided by scientific observations vis-à-vis particular scientific theories/hypotheses was discussed in considerable detail. In order to provide a precise assessment of al-Haytham's contributions to the experiment, it would be useful first to characterize briefly how observations obtained under experimental conditions would differ from observations that are made in other scientific but non-experimental contexts.

To do this, we may begin by exploring the characteristic features of an experiment. Webster's Dictionary provides the widely accepted understanding of an experiment as 'an operation or procedure carried out under controlled conditions in order to discover an unknown effect or law, to test or establish a hypothesis or to illustrate a known law'. Again, the Free Dictionary states that an experiment is 'a test under controlled conditions that is made to demonstrate a known truth, examine the validity of a hypothesis or determine the efficacy of something previously untried'.[2]

An experiment is thus, first of all, *purposeful*. It is aimed at achieving a specific goal – whether that is illustrating a known law, testing an as-yet unconfirmed hypothesis, or discovering an unknown effect.

Second, it is *deliberately conducted* by the experimenter(s) with that goal in mind. Thus, any experiment involves planning, design, and implementation. For example, let us take an experiment mentioned in Galileo's *Two Sciences* that tests the hypothesis that freely falling bodies accelerate as they fall. Here, an experiment had to be planned and designed that would enable motion to be sufficiently slow so that increases in speed at specific intervals could be measured. Galileo suggested the use of an inclined plane with a groove down which a hard brass ball could be rolled. The time the bronze ball took to roll down to cover the first quarter of the total distance, the second quarter and so on, would be determined by a water clock. A successive shortening of the time taken by the ball in later quarters would confirm his hypothesis; failure to establish this would indicate disconfirmation.

A third important feature of experiment is that it is usually conducted under *controlled* conditions. Galileo ensured for instance that the groove down which the ball was to roll was uniformly smooth and hard as possible; the parchment lining it similarly smooth. This minimized the possible effects of friction during the experiment.

Given the above considerations, it should be quite clear that observations made in experimental conditions have features that crucially distinguish them from those made outside experimental conditions. The former are achieved

---

2   Note that this study employs the broad and generic definition of 'experiment' in Webster's and the Free Dictionary, rather than the various specific definitions found in philosophical and scientific writings. This is because the very specific definitions are more likely to be author-specific and hence perhaps more controversial than the broad ones given by dictionaries (which tend to encapsulate the common understanding of the terms in question). There are also discipline-specific definitions of the term 'experiment' (e.g., in psychology, geology etc.) which would be slightly different because of the nature and concerns of the particular discipline. These too will not be considered here, as it would take the discussion too far afield.

under controlled conditions, within a context where these controlled conditions were deliberately instituted in order to achieve a particular purpose – often one involving the testing of a specific hypothesis or theory.

In this respect, experimental observations would differ from other observations that may be pertinent to science but are made in non-experimental contexts. One such example of the latter might be the inductive observation advocated by the sixteenth-century philosopher Francis Bacon. Bacon saw scientific endeavor as a matter of going about the world and making a close and careful observation of Nature. Such careful observations, he argued, could lead one to notice that certain events or features are regularly conjoined. A sufficiently substantial number of such conjunctions would allow one to inductively conclude that a law-like regularity or scientific law obtains. I shall here use inductive observation as an example to help make clear the difference in observations made under experimental and non-experimental contexts.

It is evident that observations made according to Bacon's inductive method would differ in significant ways from experimental observations. To begin with, such observations are not purposeful *in the same way* that experimental observations are. While the Baconian scientist is certainly enjoined to investigate Nature and look for regularities, she does not seek specifically to establish whether a particular outcome or a (reasonably specific) class of outcomes obtains in the way an experimental scientist does. In inductive (and other non-experimental) observations, one might not have been looking specifically for any regularity, and yet come to realize that it is present. One example of this might be the case where doctors, in the course of their clinical practice, notice that many of their patients are exhibiting an illness with a distinctive set of symptoms and an unusual trajectory. On the basis of these observations, they might then conclude that these regularities indicate that there might well be a new disease – or a mutated form of an existing disease – afflicting humans.

Such observations also differ from the observations made during an experiment in that these do not occur under controlled conditions, and in situations where the environment is deliberately and specifically arranged so that it attains a well-specified goal. The physicians mentioned above who conclude that there might well be a new disease afflicting humans may simply be going about their usual clinical work. In making their clinical diagnoses, they would not have been observing their patients' symptoms and reactions under controlled experimental conditions, nor would they have deliberately conducted – that is planned, designed and implemented – specific procedures aimed at the discovery of something new.

In sum, then, experimental observations have a number of significant features, which set them apart from inductive observations, or other kinds of
observations made in non-experimental contexts. In the next section, I look
more closely at the nature of Ibn al-Haytham's observational work, and compare this to that of his Greek predecessors. This will help make clear the nature
of his contribution to the development of the experiment.

3   Ibn al-Haytham and His Ancient Greek Predecessors

It has often been suggested that one of Ibn al-Haytham's crucial contributions
to scientific method was the emphasis he placed on empirical investigation. In
this respect, he would have been significantly different from his predecessors,
the ancient Greeks. Briffault notes of the latter:

> The Greeks systematized, generalized and theorized, but the patient
> ways of investigations, of the accumulation of positive knowledge, the
> minute methods of science and prolonged observation and experimental
> inquiry were altogether alien to the Greek temperament... What we call
> science arose in Europe as a result of a new spirit of enquiry, of new
> methods of experiment, observation, measurement, of the development
> of mathematics in a form unknown to the Greeks. That spirit and those
> methods were introduced into the European world by the Arabs. (Briffault
> 1927: 181)

Briffault here argues that the ancient Greeks were interested in theorizing
and systematization, but did little in the way of 'observation and experimental enquiry'. The 'new methods of experiment, observation and measurement'
were instead introduced by the Arabs. And of course, one of the chief contributors in this respect would have been Ibn al-Haytham.

Briffault's assertion here that the 'new methods of experiment' were brought
about by the Arabs, deserves closer inspection. As indicated in the previous
section, one of the key features of observations made during an experiment is
that they are made under controlled conditions, within a context where these
conditions are deliberately instituted in order to achieve a particular purpose.
But such a feature was certainly not unknown in earlier work of the ancient
Greeks. For example, Ptolemy, a noted Greek predecessor of Ibn al-Haytham
in the field of optics, evidently did conduct investigations that very closely
approximate the experiment as we know it. Indeed, Ibn al-Haytham himself

notes and discusses observations evidently made by Ptolemy under experimental conditions. For example, in his *Doubts about Ptolemy*, he notes:

> [Ptolemy] also says [the following] at the end of the fifth *maqala*: let three containers be made of pure and clear glass; let the first be made in the shape of a cube; let the second be cylindrical-convex; and let the third have a cylindrical-concave surface. He then says: let them be filled with water; let rulers be inserted into them and let their images be examined.
>
> Now he had shown before, that glass is more dense than water and air. From this it follows that when the ray reaches the glass from the air, it is refracted because the glass is denser than air; then when this ray reaches the water it is refracted again because it is more subtle [i.e. less dense] than glass. [Therefore,] not all the rays reaching the water will be perpendicular to the surface of the instrument, and thus pass through straight on ... And since these rays have been refracted twice, the images [of the rulers] which [Ptolemy] asserted for the rulers will become invalid.[3]

Here, evidently, Ptolemy effects a specific experimental set-up involving three containers of different shapes filled with water. Experimental controls are clearly in place – all three vessels are made of 'pure and clear glass', and the experiment was evidently set up with a specific purpose in mind (perhaps to discover some unknown effect). Admittedly, Ptolemy might not have been very good as an experimental scientist – if Ibn al-Haytham is correct, he does not get the images for the rulers right. But clearly the experimental method itself was not unknown in the world of ancient Greece.

The logical empiricists and Popper had both deemed that the most crucial role or purpose played by experiment was to confirm or disconfirm a scientific hypothesis, or to adjudicate between two competing hypotheses. But, as Sabra points out, this feature too was present in Ptolemy's *Almagest*.[4]

Sabra notes that in *Almagest* vii, Ptolemy examines the evidence for the doctrine that zodiacal stars do not change their position with respect to those that are outside the zodiac. Hipparchus had adopted the doctrine on the basis of a few and unreliable observations. Ptolemy however suggests that the doctrine will receive much stronger support if he, Ptolemy, effects certain observations and then compares them with the more reliable ones made by Hipparchus. Accordingly, he selects a number of the more readily

---

3   Sabra's "Ibn al-Haytham's Criticisms of Ptolemy's Optics," in Sabra (1994), pp. 148–149.
4   Sabra's "The Astronomical Origin of Ibn al-Haytham's Concept of Experiment," in Sabra (1994), pp. 133–36.

recognizable star configurations picked out by Hipparchus, and independently makes his own observations with respect to them. The absence of any discrepancy between his and Hipparchus' observations provides him with confirmation that the doctrine of the zodiacal stars with respect to those outside the zodiac is correct.

Once again, we find Ptolemy carrying out an experiment here. He designs and implements a specific set of procedures in order to effect a set of observations that will enable him to confirm or disconfirm whether a specific hypothesis is correct. Here too, the ancient Greeks were evidently effecting what we would call experimental observations. Further, they carried them out for the purpose (deemed crucial in pre-Kuhnian philosophy of science) of using these observations to confirm and disconfirm theories.

At this point, it might appear that claims that Ibn al-Haytham made crucial contributions to the method of the experiment are somewhat exaggerated. After all, the experimental method as we know it today was apparently known and practiced in ancient Greece, and therefore not in itself a contribution made specifically by Ibn al-Haytham. Moreover, the ancient Greeks apparently used such experiments as a means of confirming or disconfirming specific hypotheses. Here too, they employed the experiment in what we recognize (or used to, before Kuhn and Feyerabend) as the crucial context of justification of scientific theories. Thus, it would seem that Ibn al-Haytham did not bring anything obviously new to the Western world in respect of the experiment.

However, such a view would be quite unfair to Ibn al-Haytham. Despite the fact that he was not the initiator of the experimental method, or its role in the context of justification, he did indeed make crucial contributions to the development of the experiment. I will now try to outline what his contributions are.

We already know that in optics Ibn al-Haytham was the great *synthesizer*. He brought together elements of the competing intromission theory of Aristotle and extramission theory held by the likes of Ptolemy, and melded them with Galen's investigations into the structure of the eye to come up with an account of optics that was truly groundbreaking and set the investigative paths for optics in the next few centuries.[5] With respect to his contributions to the experiment, Ibn al-Haytham may be said to have been the great *systematizer*.

Here note that, while the ancient Greeks evidently did carry out activities that very closely approximate our scientific experiments, they did not have a *concept* of the experiment as a distinctive procedure. Sabra argues that Ibn al-Haytham was in fact the first to recognize and acknowledge the experiment as a

---

5   See for example, Lindberg's "The Science of Optics," in Lindberg (1978), pp. 338–368; Bala (2006), pp. 85–94.

specific and distinct procedure, different from other procedures in science. He maintains that this can be seen in the language he uses in the *Kitab al-Manazir*. The most commonly used word for 'experience' around Ibn al-Haytham's time was *tajriba*, and it would have been used to describe whatever a scientist, or proto-scientist, observed or experienced. In the *Kitab al-Manazir* however, Ibn al-Haytham consistently eschews the use of this word in describing his experimental work, instead using the Arabic term *i tibar*, and its cognates *i tabara* and *mu tabir* These terms all derive from the verb *abara*, which means 'to go through' or 'to traverse'. Significantly, as Sabra points out, the meaning of the eighth form *i tabara* is that of 'drawing an inference of one thing from another'. (Sabra 1994: 133) Clearly, then, Ibn al-Haytham recognizes that what he was doing, say in optics, involved a distinctive process in which one 'went through' or 'traversed' some set of procedures. He also recognized that such traverse involved the drawing of inferences of some sort. His description of the experimental portions of his empirical work in terms of *i tibar* and its cognates thus clearly implies he had *conceptual* recognition of the crucial features of the experiment as we know it. That is, he recognized that what he was doing was not 'mere' observation, but that it involved (a) the setting up and traversing through a set of procedures (as Galileo did when he designed, implemented and measured in his 'free-fall' experiment) and (b) some inference on the basis of what is observed (Galileo's conclusion that his free-fall hypothesis was correct). Importantly, subsequent Latin translators of Ibn al-Haytham's work rendered *i tabara* as *experimentare, i tibar* as *experimentum* and so on, and in this way introduced both the concept of experiment and the actual term 'experiment' into the scientific vocabulary of the West. Sabra sums up:

> It is not the fact that a large number of experiments are described by Ibn al-Haytham that we find remarkable in his book but rather the fact that he consciously and systematically operates with a concept of experiment which he associates with the cognates of one and the same root. (Sabra 1994: 134)

Ibn al-Haytham may not have been the first person to conduct experiments as we know them, but he was the first to recognize the procedure of experiment as a distinct and crucial approach in science, different from ordinary 'experience'. In this way, he contributed to the systematization of the scientific discipline.

Apart from this very significant contribution, there is also another way in which Ibn al-Haytham made an important advance with respect to the experiment. As I have suggested earlier, Ibn al-Haytham was not the first person to use the method of experiment, and to use experimental observations in the context of justifying a hypothesis. However, what was significant about Ibn al-Haytham was the unfailing regularity and efficacy with which he justified his each and every hypothesis through the use of experiment. Thus, to establish that straight lines exist between the surface of the eye and each point on the perceived surface of the object, he devised an experiment involving rulers and tubes. In this experiment, one looks through a cylindrical tube at an object of suitable size. Al-Haytham points out that if any given portion of the opening is covered, what is screened off would be the part that lies on a straight line with the eye and the screening body. (The ruler does the job of securing that straightness). Thus, a straight line exists between the eye and the perceived object's surface.[6]

Again, to establish that light radiates from every point of a luminous object such as a fire or lamp, he devised the following experiment. A large copper sheet is used with a circular hole in the centre. The experimenter then slides a cylindrical tube through this hole. One end of the tube is open and the other end closed but punctured to create a needle-sized aperture. In a darkened chamber, the experimenter holds a candle up to the open end of the cylinder and an opaque object up to the aperture at the other end. Only a limited amount of light from the flame thus passes though the aperture, the rest being blocked by the copper sheet.

He notes that as the candle flame moves, the light projected on the opaque object changes. When the tip of the flame is opposite the aperture, the light on the object is narrow and dim; when the center of the flame is opposite the aperture, the light on the object is wide and bright. He thus concludes that light radiates from every part of the fire.

We should also note that the above experiments did not establish only the validity of their respective hypotheses. They offered collective support for the intromission theory, which holds that vision is accomplished when light travels from the object to the eye, and collective disconfirmation for extramission theory, which held that vision is accomplished when light travels from the eye to the object.

Sabra also notes in the quote above that it was not 'the fact that a large number of experiments are described by Ibn al-Haytham that we find remarkable' but his systematic employment of a distinct concept of experiment. In point of fact, however, the large number of experiments carried out and described by Ibn al-Haytham in the *Kitab al-Manazir* are remarkable, for they point towards another important contribution that he made in respect of the experimental method. This contribution was the situating of experiment squarely as a

6   The above examples here are adapted from Steffens (2007).

necessary evidential support for the confirmation or disconfirmation of scientific theories and hypotheses. While the early Greeks may have used experiment to justify theories or hypotheses, they did not see experiment as an *essential* step in such justification. Thus, they did not always assign experiment this specific evidential role within scientific method.

On the other hand, the large number of experiments described by Ibn al-Haytham *are* remarkable because they are the manifestations of an underlying assumption on his part. He conducted a great many experiments because he was convinced that every hypothesis had to be carefully justified through experimental observations, and that this was a necessary part of the scientific enterprise. For any hypothesis that he put forward, he *had* to have an experiment that either supported or disconfirmed it – hence the great number of experiments.

In sum, then, Briffault was not far wrong when he said that 'the minute methods of science and prolonged observation and experimental inquiry' were alien to the Greeks. They might have occasionally used – as Ptolemy did – the experimental approach, but they did not see the patient and careful conduct of experiments as an *essential* feature of the scientific endeavour. This latter was Ibn al-Haytham's crucial contribution to scientific method.

At this point, however, one needs to be a little careful – al-Haytham's contributions would be crucial *only if experiments are in fact a crucial part of scientific procedure*. In recent times, it has been suggested that the role of experiment in scientific procedure may not be crucial in the way previously assumed. This is because, it has been argued, experiments do not offer independent justification for the acceptance or rejection of scientific theories. They do not do so because all experimental observations are inevitably informed or colored by the theoretical commitments of the observers in question. If this position is correct, it would certainly diminish the value of Ibn al-Haytham's contributions in respect of the experiment. There is little point in systematically conducting an experiment to 'justify' a hypothesis or theory if the experimental observations one records are already shaped, indeed infected, by the theory that one seeks to justify.

However I would like to contest such a conclusion by examining more closely Kuhn's claim that all observations are informed by the theories espoused by the observers in question. Obviously, a systematic and exhaustive attempt to assess Kuhn's detailed claims would be far beyond the scope of this study. Nor would it be possible in this space to consider developments in related areas by philosophers like Feyeraband, Lakatos, Laudan, and even Cartwright and Giere. What I hope to do instead is to indicate briefly some grounds for thinking that even if observations are shaped by the theoretical commitments of the scientific observer the experiment can still play an important role in the context of scientific procedure.

4   Ibn al-Haytham's Experimental Method and the Theory-Ladenness Thesis

As indicated above, prior to the work of Kuhn, the common consensus was that observations, and more particularly experimental observations, had their place in the scientific enterprise in that they could be appealed to in order to justify the acceptance or rejection of a scientific theory. They could also be used to choose between competing scientific theories. One example where experiment was thought to have played this role was when Arthur Eddington deliberately conducted a test between Einstein's theory of general relativity and Newtonian physics. He did this by making a special trip in 1919 to Principe Island in Africa to record the positions of stars during a solar eclipse. His observations concerning the positions of the stars agreed with the predictions of gravitational lensing made by Einstein's general theory of relativity. On the other hand, they could not be effectively accounted for under Newtonian physics. Eddington's observations were generally accepted as having provided confirmation for Einstein's theory, and evidence against Newton's theory.

In 1962 Thomas Kuhn argued in his seminal *Structure of Scientific Revolutions* (hereafter SSR) that this well-accepted account of the role of observation and experiment with respect to scientific theories was quite erroneous. Using episodes from the history of science, he pointed out that observations made by scientists were never 'objective', in the sense they were independent of the theoretical commitments of the particular observer. Thus, he argued, when Aristotle looked at swinging stones suspended by strings, he literally saw constrained fall. When Galileo looked at a similar set-up, he literally saw a pendulum. (Kuhn 1970: 121) Kuhn thus held that one could not choose between theories on the basis of crucial experiments, because experimental observations were never objective and theory-independent. However, in a paper in *The Essential Tension*, Kuhn allowed that theory choice can be made on the basis of factors such as the simplicity of the theory or its potential fruitfulness[7] although he was clear that it cannot be made on the basis of theory-independent observations – because they did not exist.

The arguments of Kuhn, if correct, would *prima facie* pose a serious threat to Ibn al-Haytham's place in history in respect of the scientific experiment. Ibn

---

7   Kuhn's "Objectivity, Value Judgment and Theory Choice" in Kuhn (1977), pp. 320–39.

al-Haytham's careful and detailed empirical work, and his careful design and implementation of experiments would *prima facie* count for nought if it turns out that the observations in those experiments were shaped by his theoretical commitments. Let us now examine a conspectus of recent arguments to the effect that it is not entirely obvious that theory-ladenness constitutes such a threat to the experimental method.

The historian of science, Peter Galison[8] notes that Kuhn's account assumes that scientific theory and scientific observation and experimentation march in lockstep with each other. He argues that this is not always the case. First, there are certain areas of contemporary science where the experimental work and theoretical work is basically carried out by different sets of scientists. One key area is that of modern physics. Given that the experimenters here are not the same individuals as the theorists, he argues that it is hard to see how theory – at least the theoretical commitments of the latter – could shape the observations of the former.

Second, Galison also notes that the work of experimentalists quite frequently span several theories or even groups of theories. In such a context, it would be hard to see how *one* particular theory could guide, shape or influence the observations of the experimenters.

For example, Galison points out that gyromagnetic experiments began before Bohr's first model of the atom, and continued during the development of the initial quantum theory, and finally that of quantum mechanics. (Galison 1987: 248) Thus it is not the case that an entire theory-cum-(theory-infected) experimental nexus stands and falls together. Experimental work and theoretical work each have in fact a separate trajectory of rise and fall. This again makes it unlikely that observation and experiment are always informed or shaped by a specific set of theoretical commitments – since they do not always march in lockstep with each other.

There are therefore a number of reasons for thinking that there can be a relative independence between experimental and theoretical research – that experimental work, and especially experimental observation – *need not* be shaped or be infected by the specific scientific theory they confirm or refute. Thus, while Kuhn might well be right that experimental observations are shaped by the observer's theoretical commitments, it is not the case that they are necessarily shaped by the theory they are designed to test. Similar arguments have been developed by Ian Hacking (1983).

The work of Galison and Ian Hacking also suggests, importantly, that there is a distinctive culture of experimental research, which is quite different in

8   The points made here are drawn from Galison (1987).

its goals, norms and concerns from theoretical research. The criteria by which experimental research is adjudged successful are different from the criteria by which theoretical research is judged. Galison notes:

> Experimental culture is grounded in expertise – the ability to eliminate kinds of background and an instinctive familiarity with the valid limits of apparatus. (Galison 1987: 248)

A good experimenter must be in the first instance a good *technician* – someone who is able to design, implement and manipulate her apparatus well and to use it to the best advantage.

A good experimenter also brings to her work certain habits of mind. One important characteristic of the good experimenter is the ability not to let her theoretical commitments – that are extraneous to the experimental situation stand in the way of her observations. In his *The Art of Scientific Investigation*, W.I.B. Beveridge writes:

> Claude Bernard considered that one should observe an experiment with an open mind *for fear that if we look only for one feature expected in view of a preconceived idea*, we will miss other things. This, he said, is one of the greatest stumbling blocks of the experimental method, because, by failing to note what has not been foreseen, a misleading observation may be made. (italics mine) (Beveridge 1950: 103)

Beveridge also notes that Darwin's son wrote of his father:

> He wished to learn as much as possible from an experiment so he did not confine himself to observing the single point to which the experiment was directed, and his power of seeing a number of things was wonderful... There was one quality of mind which seemed to be of special and extreme advantage in leading him to make discoveries. It was the power of never letting exceptions pass unnoticed. (ibid.)

Faraday, too, was someone who was well-known for not just looking for the expected, but to watch for other things.

Beveridge concludes:

> If when we are experimenting, we confine our attention to only those things we expect to see, we shall probably miss the unexpected occurrences [and] it is the exceptional phenomenon which is likely to lead to an explanation of the usual. (ibid.)

A good experimenter is therefore someone who in addition to being technically competent with her apparatus, also has a particular habit of mind – viz.,
keeping the mind open to new possibilities and alternative interpretations.
She is someone who does not just look for what she expects to see, but importantly keeps an eye out for the unusual, for what is unexpected, and which
does not fit in with the expected schema.

At this point, it would be apposite to go back to Ibn al-Haytham's own experimental work. As mentioned, ancient Greeks such as Ptolemy did do experiments. But Donald Hill distinguishes Ibn al-Haytham's work from theirs as
follows:

> One of the chief characteristics that distinguishes Ibn al-Haytham's work
> from that of his [Greek] predecessors is his rejection of the axiomatic
> approach in which postulates were assumed to be self-evident and any
> experiments were designed solely to reinforce the axioms. In contradis-
> tinction, Ibn al-Haytham was concerned predominantly with first prin-
> ciples and their justification. He regarded this as the first step in a truly
> scientific investigation. He was acutely aware of the fallibility of sense-
> perception and it is scarcely an exaggeration to say that his efforts to cir-
> cumvent this fallibility in gaining knowledge about the world was the
> generating force of his method. (Hill 1993: 72)

Hill notes that the Greeks generally conducted experiments solely to 'reinforce
their axioms'. Their goal in experimenting was merely to confirm – rather than
objectively inspect – whatever their theory propounded. This might explain
why their experimental work was not only patchy, but also sometimes shoddy.
Recall that Ibn al-Haytham expressed doubts about Ptolemy's assertions concerning the images of the rulers in his experiment involving three glass containers. It is not beyond the realm of possibility, given what Hill has said, that
these observations were marred because they were shaped by the theoretical
pre-conceptions of Ptolemy.

In contrast, Ibn al-Haytham's experimental work proved influential and
important to subsequent modern scientists precisely because it was careful,
generally free of preconceived expectations drawn from the theory under test,
and because of that, entirely replicable by others. That he had the habits of
mind requisite for a good experimenter can be seen, not only in the experimental work recorded in the *Kitab-al-Manazir*, but in his statement:

> We should distinguish the properties of particulars, and gather by induc-
> tion what pertains to the eye when vision takes place and what is found
> in the manner of sensation to be infirm, unchanging, manifest, and not

subject to doubt. After which we should ascend in our inquiry and reasonings, gradually and orderly, criticizing premises and exercising caution in regard to conclusions – our aim in all that we make subject to inspection and review being to employ justice, *not to follow prejudice*, and to take care in all that we judge and criticize that we seek the truth and *not be swayed by opinion*. We may in this way eventually come to the truth that gratifies the heart and *gradually and carefully* reach the end at which certainty appears; while *through criticism and caution* we may seize the truth that dispels disagreement and resolves doubtful matters.[9] (Emphasis mine)

Ibn al-Haytham's lasting contribution to the method of experiment was thus not just introducing the distinctive concept of experiment, but to embody, display and model the habits, methods and independent stance of the *effective* experimenter. George Sarton had called Ibn al-Haytham 'the best embodiment of the experimental spirit in the Middle Ages'. Surely the experimental spirit included the qualities of being independent, unprejudiced and careful in one's experimental investigations.

In embodying these qualities, the experimenter is arguably less likely to be influenced by theoretical commitments of the theory under test when making observations, even if it may be shaped by other background theoretical perspectives. We see this in Ibn al-Haytham himself: he was clearly not in the grip of either the prevailing intromission and extramission theories of optics, since he ultimately drew from both, and from Galen's work as well, to develop his own theory of optics.[10]

---

9   Quotation from Steffens (2007), *Ibn al-Haytham*, p. 62.
10   My discussion of Ibn al-Haytham in relation to Kuhn and Feyerabend focuses only on their connections concerning scientific experiments. Arguably, one major omission in this discussion concerns the considerable role played by scientific models in the advancement of science. This omission is a considered one: the latter half of this study examines Ibn al-Haytham's contributions to experimental method *in the context of Kuhn and Feyerabend's work*, which is largely concerned with the connections between scientific theories and experimental observations. Kuhn and Feyerabend thus accord relatively less attention to the role played by scientific models in their accounts. Insofar as this study hopes to articulate a possible rebuttal to Kuhn and Feyerabend's accounts (using Ibn al-Haytham's own work), I have responded to their accounts according to their own terms (i.e., in terms of a discussion on the connection between theory and observation).

   However one question that may come up is whether experimental observations may be 'infected' by the scientific models that the scientist employs (if not the scientific theory itself). This is a question that has to be explored but will not be done here. But a quick answer I would give here is that it is possible for the *good* or virtuous experimenter not to

Thus even if Kuhn may well be correct in claiming that there are many episodes in the history of science in which the observations of scientific experimenters were shaped by their theoretical commitments, the arguments above suggest that it in no way diminishes the significance of the experimental method, and the contributions made to it by Ibn al-Haytham.[11]

Bibliography

Bala, Arun (2006). *The Dialogue of Civilizations in the Birth of Modern Science*. New York: Palgrave Macmillan.

Beveridge, W.I.B. (1950). *The Art of Scientific Investigation*. London: Mercury Books. (Originally published by William Heinemann, 1950).

Briffault, Robert (1927). *The Making of Humanity*. Sydney: Allen and Unwin.

Crone, Robert A. (1999). *A History of Colour; the Evolution of Theories of Light and Colour*. Netherlands: Dordrecht: Springer.

Feyerabend, Paul. "An Attempt at a Realistic Interpretation of Experience," *Proceedings of the Aristotelian Society* 1958 (58): 143–170.

Galison, Peter (1987). *How Experiments End*. Chicago: University of Chicago Press.

Hacking, Ian (1983). *Representing and Intervening: Introducing Topics in the Philosophy of Natural Science*. Cambridge: Cambridge University Press.

Hill, Donald R. (1993). *Islamic Science and Engineering*. Edinburgh: Edinburgh University Press.

Kuhn, Thomas (1970). *The Structure of Scientific Revolutions* (2nd ed.). Chicago: University of Chicago Press.

——— (1977). *The Essential Tension*. Chicago: University of Chicago Press.

Lindberg, David C. (1978). *Science in the Middle Ages*. Chicago: University of Chicago Press.

Sabra, A.I. (1994). *Optics, Astronomy and Logic: Studies in Arabic Science and Philosophy*. Aldershot: Variorum.

Sarton, George (1927). *A History of Science*, vol. 2. New York: W W Norton.

Steffens, Bradley (2007). *Ibn Al-Haytham: First Scientist*. Greensboro: Morgan Reynolds.

---

succumb to having her observations shaped by the model in question. The good experimenter, in Ibn al-Haytham's own words, does not 'follow prejudice' and is not 'swayed by opinion'. She would presumably take care that her observations are infected neither by scientific theories nor scientific models.

11   I thank Ruey-lin Chen and CL Ten for their helpful independent comments on earlier versions of this study.

CHAPTER 6

Averroes and the Development of a Late Medieval Mechanical Philosophy

Henrik Lagerlund

1   Mechanism and Its Medieval Background

Mechanism or mechanical philosophy as defended by Thomas Hobbes, René Descartes and Pierre Gassendi offers a general picture of how the physical world is to be explained. It is often seen as a rival and eventual replacement for Aristotelianism. According to standard historical accounts, this general picture would dominate physical theory from the 1630's up to the middle of the eighteenth century (Gaukroger 2006: 253). There are two primary aspects to the development of mechanical philosophy in the seventeenth century: conceptions of matter and the laws of motion governing natural change. Mechanistic philosophers argued for a passive, atomist or corpuscular, view of matter. They aimed to formulate scientific laws that would capture the efficient-causal relations between these material parts. It has become standard to distinguish Aristotelianism from mechanism by reference to the concepts of quality and quantity (Gaukroger 2006, Chapter 11; Henry 2002: 69–70; Jacob 1998: 77). In my view, it is not strictly speaking a mistake doing this, but it is insufficient.

The conventional wisdom is something like the following. For Aristotle there are three states of a substance: act, potency and motion. Act and potency are linked by motion. In every case of motion the thing in question is both actualized insofar as it has already changed and potential insofar as the change has yet to be completed. Motion itself is thus a relation between act and potency. In the *Physics*, Aristotle defines it as "the fulfillment of what is potentially, as such, is motion" (201a11). Aristotle recognizes at least three different kinds of motion: a thing can be *altered* in quality, as when a body is heated; it can be *augmented* or *diminished* in quantity, as when a sprout becomes a mature plant; it can *change its place*. Change of place was called "locomotion".

According to the standard account, the rise of mechanism in the seventeenth century was a reaction to the Aristotelian conception of natural change. Quality, first of all, was banished from science. Physics is reduced to the science of mathematics, that is, pure quantity. Descartes' view of matter conforms to this. Material things are not heavy or light, wet or dry, hot or cold, but only extended. The universe is constituted by homogeneous matter without any power or activity in itself. All we know about it is that it has extension, the parts of which are all in motion relative to each other. Shape, size, and motion are the only genuine properties of bodies; all sensible qualities are to be reduced to these properties. Descartes goes so far as to insist that matter *is* quantity. Of the three kinds of motion in Aristotle's physics, only locomotion survives: alteration and augmentation are to be explained in terms of bodies changing their locations (or, in some cases, fusing or dividing).

In *Physiologia*, Dennis Des Chene argues against this naive distinction according to which Aristotelian physics is qualitative and the new physics is quantitative. In the sixteenth century, and earlier, Aristotelian physics is not merely qualitative. The new physics, moreover, despite its claims to have "geometrized" nature, in fact often takes over without change qualitative explanations from its predecessor, supposing or promising that references to quality will be discharged in an eventual reduction. The problems De Chene introduces into the traditional picture can be pushed further back all the way to fourteenth-century discussions of the nature of substance, power, and quantity. It is there we should look for the background to sixteenth- and seventeenth-century views on matter and the mathematization of nature.

Aristotelian explanations also invoked four kinds of cause: efficient, material, formal and final. Not until these causes are given is the explanation of a phenomenon complete, according to him. One key difference between Aristotelian philosophy and the new mechanism was that in the latter only efficient causes were recognized or required. The final cause or end for which natural agents were supposed to act came in for particularly harsh criticism: the Aristotelian, in attributing ends to natural changes, was said to have confusedly bestowed minds on natural agents. Spinoza, for example, argues in the appendix to *Ethics* I that final causality turns nature upside down and puts the effect before the cause. Suppose, for example, that a stone, in falling off a building, kills someone. According to Spinoza, the Aristotelian must hold that the outcome – death – is the *end* toward which the efficient cause – the stone – acts when it falls. The outcome is then held to be the cause, and the stone's action, or the stone itself, an effect. This is what was meant by turning nature upside down and only minds could act for an end. Descartes, for his part, argued that since we cannot know God's intentions natural philosophy has no business looking for ends in nature. Instead the natural philosopher should merely try to discover the laws governing motion.

Scholars of early modern philosophy sometimes fail to notice – their authors often don't mention it – that the critique of final causes, and the very arguments themselves adduced by Descartes, Spinoza, and other mechanical philosophers, have precedents in the fourteenth and fifteenth centuries. John Buridan in the first half of the fourteenth takes a position as radical as Spinoza's, giving primacy to the efficient cause (Lagerlund 2011).

It is part of the traditional view of the development of mechanical philosophy and modern science that internal changes in physics had the principal role in leading thinkers like Galileo and Descartes to question and reject Aristotelian explanations. This was the foundation of their rejection of Aristotelian metaphysics of nature, that is, of form and matter, substance, actuality and potentiality, quality, and the four causes. (Lindberg 1992: 361; Henry 2002: 71; Jacob 1998, Chapter 2) Such a treatment of early modern science is not entirely accurate; instead certain *metaphysical* presuppositions had to be changed for the changes in physics to be possible. A conceptual shift was needed before the scientific shift could take place. This conceptual shift was made in the early fourteenth century by thinkers like William Ockham and John Buridan. It has been known for some time that these thinkers played an important role in the development of science (Duhem 1908, Maier 1955, Des Chene 1996, Funkenstein 1989). Nevertheless it is still not clear precisely how or in what way.

There are three major conceptual changes that took place in the early fourteenth century, all of them crucial for developments in the seventeenth century. The first of these concerns power or activity, the second natural laws and teleology, and the third substance and quantity. The remarkable similarities between Ockham and Buridan and certain early modern thinkers like Descartes is not a coincidence.

One central criticism of Aristotelian physics in early modern philosophy concerns power. The Aristotelian view of nature attributes real powers or activities to nature itself. Any motion or change on this view of nature is an actualization of some power or potency in the substance itself. Descartes and other mechanistic thinkers denied that there are any active powers in nature. According to Descartes, only the human mind (for him not part of nature) possesses an active power to effect change; the natural world is entirely passive, the things within it mere arrangements of corpuscles. These corpuscles of matter were unobservable and thought to have only some geometrical properties, like size, motion, shape, etc.

A very similar view of nature was developed by Ockham in the early fourteenth century. He wanted in general to reduce his ontological commitments. In aid of this he eliminated powers from among the kinds of things existing in the world. Starting from his predecessor John Duns Scotus's distinction between natural and rational powers, Ockham claimed that there are no powers in nature that arise from nature itself. All power is bestowed on it by God or some other active power like the human mind. Ockham, like Descartes, was a dualist with respect to mind and body. He attributes power or activity to minds alone. The mind has the "rational" power to will. Ockham's elimination of powers from all natural substances is of absolute importance for the development of a modern view of nature.

Final causality is absolutely fundamental for Aristotelian physics, and as Aristotle explains in the *Physics*, if there is no final causality then everything happens by chance. The critique of final causality was therefore a cornerstone of the mechanical philosophers attack on Aristotelianism. But Ockham and Buridan had already in the fourteenth century fundamentally reshaped the discussion and it's their debate that effected future generations, as can be seen in Suarez's treatment of final causality (Schmaltz 2008 and Akerlund 2011).

Ockham made it very clear in his *Quodlibetal Questions* IV, q. 1, what he thinks final causality is:

> I claim that the causality of an end is nothing other than its being loved and desired efficaciously by an agent, so that the effect is brought about because of the thing that is loved.

The object of someone's love is the end of that person's actions, which are caused efficaciously for the sake of that end. The efficient cause of someone's action is the will, but the final cause is the object loved. He is very clear further on in the same *Quodlibetal Question* that there is no final causality in nature and draws a sharp distinction between free agents and natural agents. The final cause is in the mind and not in nature on this view.

Buridan works out the details of Ockham's suggestion. His rethinking of final causality is motivated by the very problem that Spinoza would highlight three hundred years later. Buridan argues that "every cause is naturally prior to the caused thing /.../ but the end is not naturally prior ..." The problem is thus how something that is posterior to its effect can be a cause. To do so it must act backwards in time, since causes are naturally thought about as that which brings something about. The argument that worries him most seems however to be the following. Take the traditional Aristotelian picture that the cause of the doctor's prescription of a certain medicine is the health of the patient. By analogy, then, it must follow that the cause of God's action of creation is the world being created. But this implies that things inferior to God are causes of His actions, which is clearly absurd. Buridan takes this to be a knock-down argument against final causality.

On Buridan's view the final cause is instead internal to us as our desire or reason for acting. He hence reduces all final causality both in nature and in mind to efficient causality. On such a view the argument that seems to have worried him the most, namely the argument that the world is the cause of God creating it, is not at all a problem. The world is not the cause of anything, since the world only exists as an intentional object in the mind of God and as such it is an efficient cause of God's act of creation (Lagerlund 2011).

Buridan also addresses the view present in Aristotle that everything will happen by chance if there is no final causality. He says the same thing as the seventeenth-century mechanical philosophers, namely that efficient causes are sufficient. The regularity according to which one judges whether a natural action turns out as it should is not the end, but the law according to which it is accomplished and according to which the same cause always yields, *ceteris paribus*, the same effect (Des Chene 1996).

The fourteenth century also saw some radical changes in the way substance was conceptualized. Ockham challenged the Aristotelian or Thomistic way of thinking about material substance and body as he systematically rethought metaphysics. According to Aquinas, a substance had no parts that are prior to it. A composite substance, an animal or a human being for example, comes to be out of another substance, but only what they call prime matter remains the same during this generative process and prime matter has no existence on its own. Since matter was their principle of individuation, form could have no existence before its union with matter. Hence nothing in an individual composite substance pre-exists its existence in nature.

To insist, as Ockham, that a substance is nothing but its parts is contrary to Aquinas who held that although substances have integral parts these parts depend ontologically on the whole of which they are parts. Each part of a substance is actual and not dependent on anything to make them actual, Ockham argues contrary to Aquinas.

Every substantial form in nature (other than the human intellectual soul) is extended and composed of parts, according to Ockham. Hence all the properties of the whole form are going to be derivative upon its parts and furthermore all the properties of the whole composite substance will be derivative of the properties of the parts of the form and the matter. Since the integral parts of the form parallel exactly the integral parts of the matter, and since the matter is something existing in its own right, the view defended by Ockham is a fraction away from the abolition of the whole distinction between form and matter. His view appears similar to the later seventeenth century view of substance wherein the properties of the parts of the material body are all properties which have the actual material parts of the composite as their subjects.

In his metaphysics, Ockham tries very hard to reduce the Aristotelian categories to only substance and quality. Buridan immediately challenged this, even though he agreed with Ockham's methodological principle that one should not postulate entities beyond necessity. Buridan argued, however, that such a postulate was necessary when it comes to quantity – extension and magnitude, must be separate entities. On Ockham's view all extended things are simply their extensions. One of his arguments for this was that not even God could make an extension distinct from a substance. Buridan's arguments to the contrary are derived from physics (Maier 1955).

As the debate between Ockham and Buridan was taken over by later fourteenth and fifteenth century thinkers, the importance of the concept of quantity for physical explanations generally became increasingly evident. But much more needs to be explored to understand this trend and its impact on the seventeenth century's rise of mechanical philosophy.

The ideas of Ockham and Buridan were naturally controversial. It took centuries of debate to come to grips with the consequences of their ideas and it is only by beginning to work through the details of the works of their followers in the fifteenth and sixteenth centuries that we can come to know exactly how influential they were and how their ideas transformed thinking about nature.[1]

2   Matter in the Fourteenth Century and Its Background in Averroes

In the second part of this article, I would like to look a little bit closer at a particular aspect of the view of matter that developed in the footsteps of Ockham and Buridan. The notion of matter found here plays a crucial role in the development of a new physics. In this discussion the notion of *minima materia*, or *minima naturalia* as it was also called, played a distinctive role. It is, as will become clear, not exactly the same notion of *minima materia* that can be found earlier in the Latin Aristotelian tradition. (Murdoch 2001). Albert of Saxony, a colleague of Buridan in Paris, writes in Question 10 of his *Physics* commentary about generation and successive development of bodies (152):

> Forms of donkeys or some other heterogeneous [body] other than a human being are generated successively.... One part of the form of a donkey is first educed from one part of the matter and after that another part of the form from another part of the matter. It follows that the form of a donkey is generated successively, and this goes for any other heterogeneous body other than a human being.

---

1   This part of my article is based on a project application that I submitted to SSHRC (Social Science and Humanities Research Council) in Canada. I am indebted to Dennis Des Chene, Benjamin Hill and Calvin Normore for their help in developing these ideas.

According to this view, forms of nonhuman substances are educed (*educere*) from matter.[2] As the text explains, this is a successive process, so parts of the form are educed from parts of the matter, and another part from another part of the matter and so on. Both Albert and John Marsilius of Inghen (another late fourteenth century thinker in this tradition) comment on this (John Marsilius of Inghen 1518, I, q. 13, f. Xiiiva):

> I posit this conclusion that a minimal matter is granted under which a heterogeneous mixed body can be generated.

Albert also says (Albert of Saxony 1999, I, q. 10, 153):

> The third conclusion is that a minimal matter is granted by which power the whole form of a donkey can be educed or a form sufficient for which the composition of the form and the matter is said to be a donkey.

On their view, then, every bit of extended matter is a quantity, and what these authors are insisting on is that there is a minimal boundary of the quantity of matter that can generate, that is, develop, into something with a certain form. A *minima materia* is, on this view, a minimal quantity of matter below which it cannot generate. These are not atoms or indivisibles, since they are in an Aristotelian fashion, divisible, but perhaps they are best likened to what Descartes called corpuscles. They are the units of physical interaction and change. It is also worth noticing that a mere mechanical cutting out of some adequate quantity of matter ought to be able to generate a substance.

The doctrine of *minima naturalia* was developed from comments made by Aristotle in the *Physics* and became a well-known theory of scholastic physics (Murdoch 2001). It is important to note, however, that the view developed in the footsteps of Ockham and Buridan presented here is different. In his 2004 book, T. Holden draws a distinction between what he calls (i) a potential parts doctrine and (ii) an actual parts doctrine (Holden 2004: 18). According to (i), the parts that a body can be divided into are not distinct prior to them being divided. The traditional Aristotelian or scholastic theory of *minima naturalia* is such a theory. The view developed by these fourteenth century thinkers is not such a theory, but the second kind of theory where the *minima naturalia* are actual parts.

---

2 As mentioned already thinkers like Ockham and Buridan thought humans have minds that are not part of nature, hence such entities are not generated, but created and immaterial.

In a recent and very interesting book, Ruth Glasner argues that Averroes develops a theory of matter that seems at least superficially similar to the one mentioned above (Glasner 2009). She calls this view Averroes's Aristotelian atomism[3] and she describes it in the following way:

> Averroes' theory of *minima naturalia* is a theory of actual and essential parts and, as such, it bridges the gap between the two opposing systems, the Aristotelian and the atomistic. (159)

In his discussion of motion in Book VII of all three of his commentaries on the *Physics* of Aristotle, Averroes introduces a distinction between (1) first-moved part and (2) not-first-moved part. (1) is that which is not moved due to a part of it being moved essentially and (2) is that which is moved as a whole because of a part of it is moved essentially. Hence, (2) is moved by (1). He explains in his short commentary what a first-moved part is:

> The first-moved, that is, that which is moved not due to a part of it that is moved, is that which is moved essentially. In these simple bodies, which are the cause of doubt, it is the minimal possible magnitude of fire that moves upwards or the minimal possible magnitude of earth that moves downwards, because the moved [part] of earth and fire that is so described is a first-moved because this movement cannot occur to a part of it, for a smaller part of fire cannot exist, because the magnitude of existing things are limited. (114.1–9)[4]

This first-moved part is somewhat similar to what was later called a *minimum naturale*. It is a unit that carries the essential motion of a body, and as with later Latin thinkers this unit is a form/matter composite. Averroes, hence, it seems develops a notion of matter, or *minima materia*, that comes very close to the one we have seen reported by Albert of Saxony.

---

3 For a critical review of Glasner's book see Trifogli (2010).

4 This is quoted from Glasner (2009), p. 154. A similar view is expressed in the Latin translation of his major commentary on the *Physics*: "Et necesse est hic esse prima mota, quia corpora naturalia non dividuntur in infinitum in eo, quod sunt corpora naturalia. Verbi gratia quoniam primum motum in igne est minima pars, que potest esse ignis in actu, et similiter primum motum caloris naturalis animalium est esset minima pars, que potest movere illud animal." (Averroes, Physics, VII, cap. 2, 307va.)

The textual sources of Averroes's highly original interpretation of Aristotle is somewhat unclear, since the text situation of Averroes's various *Physics* commentaries is quite messy (Glasner 2009: 10–20). It is also unclear how and if it had any influence on the Latin tradition. Averroes's particular view outlined here is, however, mentioned by both John of Jandun and John Buridan in their respective *Physics* commentaries. The question that deals with this in John of Jandun's commentary starts with an interesting argument. It is really an argument against infinite divisibility of material things, since if everything material is infinitely divisible, then there cannot be a first part.[5] There seems hence to be a conflict between the idea that there is a first part and that everything material is infinitely divisible. It seems possible, of course, that even if there is no first part there can be a first moved part. Averroes seems to assume, at least on Glasner's interpretation, that the lack of a first part also rules out a first moved part. The way Albert of Saxony might put this is that only when a sufficient quantity of matter is present to generate a form do we have a part that can move and act to form further things. Averroes agrees with this to some extent, namely that the first moving part is a unit of some matter and form. The way Jandun formulates Averroes position he assumes that the first moved part is also indivisible.[6]

Buridan also mentions Averroes view in question 13 of the first book of his commentary.[7] He adds that even though there must be a first moved part, God can nevertheless by his absolute power divide this part. By using the distinction

---

5   "Tunc dico ad questionem: primo quod si accipiatur motus pro fluxu continuo sic non est in ipso primum mutatum esse intelligendo per primum mutatum esse primam partem illius fluxus continui sicut videtur Aristotelem intelligere. Et hoc est facile videre, quia in illo quod est divisibile in infinitum non est pars prima. Hec est manifesta, sed quodlibet continuum est divisibile in infinitum. Unde iste fluxus continuus qui dicitur motus secundum famosiorem acceptionem est divisibilis in infinitum sicut et tempus mensurans ipsium ut patet in isto sexto quare et ceter." (John of Jandun, *Physics*, VII, fol. 106va.)

6   "Unde non videtur reliqui quod ipse intelligat aliud per illam primam partem nisi primum indivisibile a quo initiatur motius, et huiusmodi indivisibile potest vocari pars prima inquantum se habet quodammodo ad modum forme. Et hoc forte inquit Commentator in suis verbis. Ubi dicit: prima pars in actu idest prima pars que se habet ad modum actus et forme. Sicut enim res non est antequam forma fit et statim forma adveniente res habet esse, sic ante illud indivisibile non erat ille motus sic statim post illud indivisibile sequitur motus." (Ibid., fol. 107rb.)

7   "Ex istis dictis apparet fuisse intention Aristotelis et Commentatoris quod in alteration, generatione, et cossuptione est dare primam partem que tota simul generatur, corrumpitur vel alteratur et non mediates prius." (Buridan, *Physics*, I, q. 13, fol. xvii rb.)

between God's absolute and ordained power Buridan can hence maintain that there is a first part at the same time as he holds on to the Aristotelian view that everything material is infinitely divisible.[8]

3   Conclusion

It is clear that Averroes's very original interpretation of Aristotle was known and commented on in the fourteenth century. It was known, for example to Buridan as we have seen, when he developed his view of matter and material substances. It seems clear that Averroes's view, which in the thirteenth century was highly unorthodox as an interpretation of Aristotle, certainly fitted much better in the fourteenth century. This becomes clear if we connect the idea of *minima naturalia*, as understood by Averroes and his fourteenth century commentators, as actual parts with Ockham's view of substance mentioned earlier. The *minima naturlia* can then play the role of the actual parts that Ockham thinks make up a substance. Such a view of substance looks very similar to the views of mechanical philosophers like Descartes and also Hobbes, although in his case the actual parts are atoms. If we then put this view in the context of all the other changes and novel ideas developed in the fourteenth century, the nominalist or *via moderna* tradition developed by Ockham and Buridan has to be seen as part of or an early version of mechanism. The way this comes together also illustrates what I mentioned earlier that the metaphysical ideas of Ockham plow the way for physics and how the physics was made to fit a new metaphysics, and in this way the metaphysical changes set new limits for physics. I have here tried to illustrate that such a shift seems to be going on in the fourteenth century. However, much more research is needed to track the influences of the fourteenth century developments on the mechanical philosophy of the seventeenth century, as well as, Averroes influence on the views of matter and material substance of the fourteenth century.

---

8 "Prima conclusion est quod inter omnia corpora actu seorsum existentia sine continuation unius ad alterum est dare aliquod minimum, scilicet quo nullum aliorum est minus, quia licet in infinitum continuum sit divisibile, tamen non est possibile quod sit divisum in infinitum. Ideo corpora sic actu seorsum existential sunt finite multitudinis et non infinite et inter talia esset devenire ad illud quo nullum aliorum est maius et ad illud quo nullum aliorum est minus et forte sic intendebat Aristoteles quod? est dare minimam carnem et hoc sufficiebat sibi contra Anaxagoram." (Ibid., fol. xvi vb.)

Bibliography

Akerlund, E. (2011). *Nisi temere agat: Francisco Suarez on Final Causes and Final Causality*. Uppsala: Uppsala University.

Albert of Saxony (1999). Expositio et Quaestiones in Aristotelis Physicam ad Albertum de Saxonia Attributae, II. Ed. B. Patar. Louvain: Editiones Peeters.

Aristotle (1984). *The Complete Works of Aristotle*. 2 vols, Jonathan Barnes (ed.). Princeton: Princeton University Press.

Buridan, Jean (1509). *Subtilissimae Quaestiones super octo Physicorum libros Aristotelis*. Paris, reprinted as *Kommentar zur Aristotelischen Physik*. Frankfurt: Minerva, 1964.

Des Chene, D. (1996). *Physiologia. Natural Philosophy in Late Aristotelian and Cartesian Thought*. Ithaca: Cornell University Press.

Duhem, P. (1908). *Sur la Notion de Théorie physique de Platon a Galilée*. Paris: Sorbonne.

Funkenstein, A. (1989). *Theology and the Scientific Imagination from the Middle Ages to the Sixteenth Century*. Cambridge, Mass: Princeton University Press.

Gaukroger, S. (2006). *The Emergence of a Scientific Culture*. Oxford: Clarendon Press.

Glasner, R. "Ibn Rushd's Theory of Minima Naturalia," *Arabic Sciences and Philosophy: A Historical Journal* (2001) 11(1): 9–26.

——— (2009). *Averroes' Physics*. Oxford: Clarendon Press.

Henry, J. (2002). *The Scientific Revolution and the Origins of Modern Science*. New York: Palgrave.

Holden, T. (2004). *The Architecture of Matter: Galileo to Kant*. Oxford: Oxford University Press.

Jacob, J.R. (1998). *The Scientific Revolution: Aspirations and Achievements 1500–1700*. Atlantic Highland N.J.: Humanities Press.

Lagerlund, H. "The Incompatibility of Efficient and Final Causality: A 'New' Mind-Body Problem," *British Journal for the History of Philosophy*, (2011) 19(4): 587–603.

——— (2012). "Material Substance," in J. Marenbon (ed.), *Oxford Handbook of Medieval Philosophy*. Oxford: Oxford University Press, pp. 468–485.

Lindberg, D. (1992). *The Beginnings of Western Science*. Chicago: University of Chicago Press.

Maier, A. (1955). Metaphysische Hintergrunde der spatscholastischen Naturphilosophie, Studien zur Naturphilosophie der Spatscholastik. Bd. IV, Storia e Letteratura, Roma.

Murdoch, J.E. (2001). "The Medieval and Renaissance Tradition of Minima Naturalia," in C. Lüthy, J.E. Murdoch and W.R. Newman (eds.), *Late Medieval and Early Modern Corpuscular Matter Theories*. Boston: Brill.

Ockham, William of. (1991). *Quodlibetal Questions*. Translated by Alfred J. Freddoso and Francis E. Kelly. New Haven, Conn.: Yale University Press.

Schmaltz, T. (2008). *Descartes on Causation*. New York: Oxford University Press.

Trifogli, C. Review of *Averroes' Physics: A Turning Point in Medieval Natural Philosophy* by Ruth Glasner, *Aestimatio* (2010) 7: 79–89.

CHAPTER 7

Barbarous Algebra, Inferred Axioms: Indic Rhythms and Echoes in the Rise of Western Exact Science

Roddam Narasimha

Occidental mathematics has in past centuries broken away from the Greek view and followed a course which seems to have originated in India and which has been transmitted, with additions, to us by the Arabs; in it the concept of number appears as logically prior to the concepts of geometry.
HERMANN WEYL (1928)

1   Evolution of the Indic Number-Centric System

The above statement by one of the doyens of the then-new theory of quantum mechanics was made at the time of a turning point in the history of modern physics, and appears in the Preface of Weyl's pioneering book on *The Theory of Groups and Quantum Mechanics* (1928 German edition).[1] It is the purpose of the present essay to trace, in broad outline, the 'course' followed by western mathematics from its Indic origins. The focus will be on conceptual and epistemological issues rather than on the details of transmission of

---

1   This essay is a sequel to the paper I read at the last Institute of Southeast Asian Studies meeting in Singapore on the subject (Narasimha 2012). That paper presented an overview of Indic thought on reliable knowledge, as embodied in the "pramana" theories of the classical Indic philosophical systems; and then compared it with the deductionist logic that was the hallmark of the Greek concept of "proofs". The essay discussed briefly the relevance of these views to the science of today. Following the suggestion of Arun Bala, the present paper now provides a more detailed analysis of some landmarks in the "past centuries" of Weyl's assertion as well as more recently – landmarks that exhibit the influence of Eastern ideas on the rise of science in Europe beginning around the 16th century. I must confess that I am not a historiographer, in particular of European scientific literature, and shall therefore not venture into questions of priority and transmission in general. Instead I shall note that the different cultures of the Eurasian super-continent moved along different trajectories over the millennia, but nevertheless were at the very least in some loose contact with each other. So exchange of ideas was bound to take place and has often done so, sometimes directly and at others by some informal, random process of diffusion.

© KONINKLIJKE BRILL NV, LEIDEN, 2016 | DOI 10.1163/9789004264199_009

specific techniques or tools. Incidentally Weyl's work on group theory led him to conclude that "the present trend in mathematics is clearly in the direction of a return to the Greek stand-point". This view can be debated, but will be considered only towards the end of the essay.

Among the first eastern rhythms to be felt strongly in Europe was the Indian numeral system. The story of its spread across West Asia, Italy, north Africa, and Spain to the rest of Europe has now been well-told several times (Ifrah 1998), and it is unnecessary to repeat it here. The system has eventually covered the whole globe, being one of the few such, perhaps the only one, to be so widely accepted and used; no other symbology – even in the letters, the arts or religion – can match the universal appeal of the numeral system, or of the special kind of mathematics it triggered in Europe and elsewhere that developed during the past four centuries. Although this wide acceptance took the best part of a millennium, its spread has been inexorable. The phenomenon was therefore not so much a rhythm as a murmur that slowly created a new language.

The profound effect of a system of expressing numbers where the value of a symbol depended on its position, and every conceivable number could be expressed with only ten symbols, cannot be exaggerated. The wave of Indo-Arabic algorization reached the consciousness of Europe only by about the 15th century. It showed the world that there was a notation that made it easy to think of and manipulate large numbers wherever necessary. [Incidentally Indic civilization has reveled in large numbers for more than three millennia.] It made Francis Bacon (1561–1626), the philosophical father of modern science, write:

> And no one should be afraid of multiplicity or of fractions. For in numerical calculations one would as easily posit or think of a thousand as of one, or of a thousandth part as easily as of a whole.
>
> (Novum Organum 2 viii).

So Europe got rid of its fear of large numbers. This was the first step in defining a new mathematics in the West.

But the real achievement of the new system was the computational revolution it brought about. This was not so much a matter of speed of arithmetical calculation, for the abacus permitted high speeds too for trained fingers. But the abacus was a mechanical device that did not lead to that measure of abstraction that the zero / decimal / place-value system did. The revolution is evident in the mathematical astronomy of the *Aryabhaṭīya* of Aryabhata (composed 499, Shukla and Sarma 1976), which is basically a collection of some sixty algorithms for astronomical predictions – terse instructions in verse on the calculations to be performed. It is not clear whether the new numeral system was already in vogue in Aryabhata's time, as his work was written in verse; but definite signs of it (e.g. the expansion of large numbers in powers of 10) are already there in his work. There was also the ingenious syllabic system of expressing numbers that he introduced for presenting a table of sines in verse (Narasimha 2001). It was certainly in use by the 7th century, very probably much earlier.

The computational revolution that followed in succeeding centuries was characterized by the rapid development of new numerical procedures, algorithms, algebra, approximations, equational statements, and eventually ideas that could handle even the infinitesimals that led to the calculus. From a modern view point, the interesting thing about the *Aryabhatiya* is that several physical phenomena and effects are actually mentioned in between its numerous algorithms. For example the point is made that what is important is relative motion, with the indication that that is what is being treated in the text; thus it is the relative velocity between the sun and the earth that is the object of interest. The example is given of how for an observer in a boat sailing down a river the trees on the banks will appear to be moving in the opposite direction. Based on this argument of what came to be known as Galilean relativity, the algorithms do not need commitment to a geo- or helio-centric system, but permit Aryabhata to make the proposal that the earth is spinning on its axis and this could be what makes for the day-night diurnal cycle. Similarly eclipses are seen as purely a matter of shadows, and the fact that the earth is round is seen as an immediate consequence of the circular arc that advances across the disc of the moon during a lunar eclipse. Even more intriguing is the fact that these important physical insights often appear as interesting sidelights of – or inferences from – the success of the algorithms that the text is chiefly describing. The insights are not presented as major achievements or discoveries.

Although Aryabhata was criticized for attributing eclipses to shadows by later astronomer-mathematicians such as Brahmagupta (628), not to mention more conservative puranic scholars (see Minkowski 2004, Narasimha 2007), one gets the impression that the enduring influence of Aryabhata on Indian astronomy over more than a thousand years was largely due to the short, novel and effective algorithms that he created and assembled in his text. This is shown by the way that many traditional Indic almanacs (*pancangas*), whose authors to this day present data on eclipses in terms of the mythological *rahu-ketu* picture, will nevertheless advertise their work as following Aryabhata; and they often quote the phrase *drg-ganita* as underlying their predictions. The word *drg-ganita* here stands for seeing and computing; achieving agreement between the two was the primary objective, and the underlying concept of what I have called computational positivism.[2]

The special characteristics of the Indic concept of mathematics are probably best seen in the celebrated *Siddhanta-siromani* (*The Siddhantic Crest-jewel*) collection of Bhaskara II (b 1141) composed in 1150. This text, which represents a culmination of several centuries' effort on the development of algebra, is actually a quartet, consisting of the *Lilavati*, the *Bija-ganita, Gola-adhyaya* and the *Siromani* itself. These works respectively treat arithmetic, algebra, the mathematics of the sphere and astronomy. *Lilavati* is a pleasant and playful book with plenty of exercises set against a background of poetic imagination. *Bija-ganita* is defined as the ganita (literally reckoning, calculation or computation, but often used for the more general concept of mathematics, just as the Greeks used geometry for mathematics) that handles unknown variables (*avyakta-ganita*; bija is literally seed, presumably with a potential to yield fruit, i.e. problem solution). Bhaskara sings praises of the power of algebra, of the joy that the reader will get from his algorithms, and of how the reason for using algebra is the clarity (*sphuta*) that it will bring to mathematics. *Gola-adhyaya* handles the geometry of the sphere, especially as required in astronomy. In addition to the astronomy itself the *Siromani* introduces the concepts of derivative and integral, the solution of indeterminate quadratic equations, and the careful evaluation of a variety of corrections required in predictive astronomy (Colebrook 1817).

Equations

The same is the case with equations, which deserve special consideration as their influence on the development of exact sciences has been profound. In Indic mathematics symbols, mathematical expressions and algebraic equations make their formal appearance in the *Brahma-sphuta-siddhanta* (Bss) of Brahmagupta (628 CE, Colebrook 1817), and later in the commentary thereon by Pruthudakaswamy (8th century). There was also the Bakhshali manuscript (BM), discovered in 1881 by a farmer in a village near Peshawar. The date of the text has been the subject of much debate, and the proposals made vary from 2nd to 13th century. Hayashi (1995) suggests the date of the work as the 7th century, in the century that also saw the appearance of the Bss; others see BM as dating back to the 4th century, before the classical period of Indic mathematics which may be said to have been heralded by the *Aryabhatiya* of Aryabhata in 499. Based on the language of the text and its relatively less formal and

---

2   This is discussed at length in Narasimha 2003.

sophisticated practices in the use of mathematical symbols, it seems that an earlier date is more likely.

The notation for the unknown in the Bakhshali manuscript is just a dot, even when there is more than one in the equation; i.e. any unknown variable is indicated by a dot. Later practice uses *ya*, first syllable of the word *yavat-tavat*, which means 'as much as' or 'value as required' (for the equation to be satisfied); if there is more than one unknown the first letters of the words for various colours are used; for example *ka* for *kalaka*/black, *ni* for *nila*/blue, lo for *lohita*/red and so on.

The usual arithmetical operations are generally denoted by the first syllable of the Sanskrit word for the operation, and *follow* the quantity operated upon. Thus *yu* stands for *yuta*, 'added'; so AB *yu* is A+B. Subtraction in denoted by *ksa*, for *ksaya*, or the + sign that indicates addition today; thus AB + in Sanskrit stands for today's A–B. Similarly *gu* for multiplication, *bha* for division.

The usual way in the Brahmagupta/Prthudakasvami (B-P) system of representing equations was to write the two (equal) sides of the equation one below the other, with equal powers of the unknown aligned vertically. For example, the equation that would today be written as

$$ax^2 + bx + c = dx^2 + e\,x + f \tag{1}$$

would be represented in Sanskrit by

$$
\begin{array}{llllll}
ya & va & a & ya & b & ru & c \\
ya & va & d & ya & e & ru & f
\end{array}
\tag{2}
$$

Here *ya* stands for *x*, va (= *varga*) for the square (so *ya* va is x^2), ru (= *rupa*) for the pure number *c*. (I am here adopting the convenient modern convention of printing unknown variables in italics (as in *x*, *y* etc.), and letters indicating operations on them in roman (as in sin *x*, exp *y*). If dx^2 did not appear on the right hand side of (1), the lower row in (2) would start with *ya* va o. A linear two-variable equation

$$ax + by = c \tag{3}$$

would be written

$$
\begin{array}{llllll}
ya & a & ka & b & ru & o \\
ya & o & ka & o & ru & c.
\end{array}
\tag{4}
$$

The system was sufficiently flexible to accommodate polynomial equations in multiple variables.

Let us briefly compare such Indic systems with a well-known European example, from the work of Lua Pacioli. His book *Summa de arithmetica geometria proportioni et proportionalita*, published in 1494 in Venice (second edition 1523) some seven to eight centuries after the B-P work in India, served as a common introduction to mathematics in Europe at the time (Cajori 1928). Pacioli often uses (as in India) the first letter or syllable of a word as a symbol for a mathematical object. Thus p˜ stands for (today's) plus, m˜ for minus; *co* (from *cosa*) is an unknown quantity (rather like x today), *ce* (for *censo*) for x^2, *cu* (for *cubo*) for x^3 etc. A special symbol has *five* different uses, indicating higher powers, roots, powers as well as roots of numbers, a mark for the unknown variable and a fraction. Equality was indicated by a dash or sequence of shorter dashes.

An example from Cajori will illustrate similarities and differences between Paciolo's notation and the B-P system. The equation

$$x^2 - y^2 = 36$$

is written by Paciolo as

1.ce.m˜.1.ce.de.β _____ 36

where the symbols stand respectively for $1.x^2$, minus, 1.,square, of, y equals, 36. (Here β stands for a symbol in the original that cannot be reproduced in print; a misprint in the original, pointed out by Cajori (p.111), has been corrected.) In the BP system the same equation would be written in the form

va *ka* 1   va *ni* i   ru o
va *ka* o   va *ni* o   ru 36

where *ka* and *ni* stand for the two unknowns x and y, and the dot above the second unit, 1, indicates −1.

I do not know whether the Indian system travelled to Europe over the centuries (which it could have), but there are obvious and interesting similarities, and some differences, between the two. The use of syllables for indicating unknowns and squaring operations is common; but the notation for an equalizing statement has different formats, with the Indian appearing more systematic and uniform, but including (for the sake of clarity) terms (like those involving o above) which in a better system might be unnecessary.

Calculus

Calculus was briefly mentioned above but it is now clear that a variety of results considered as part of that subject were known and used in India long before Newton and Leibnitz. The concept of the derivative (*tatkalika* = instantaneous) was already there about a thousand years earlier. Finite differences had been used for a long time: Aryabhata's table of sines actually lists the first differences rather than the function itself (Narasimha 2001). Brahmagupta used the second order finite difference interpolation scheme attributed later in Europe to Newton and Stirling; the Newton-Gauss scheme was used by the Kerala mathematicians. The *Surya-siddhanta* (600–1000) gives a simple rule for calculating the sine of an angle (Playfair 1798) which is really a second-order forward difference scheme for solving the second order ordinary differential equation that governs the sine (Narasimha 2008). Madhava presented infinite series for trigonometric functions (Sarma et al. 2008), of the kind later rediscovered by Maclaurin (1698–1746). In particular he found for the arctan the infinite series known after Gregory, and used it for calculating *pi* to 11 decimal places. Incidentally, in deriving the series he makes use of an asymptotic approximation for any large n in an expression involving integer n (Narasimha 2008).

Summary

Three things stand out from this brief account. First, it is clear that, in India, arithmetic, algebra, equations and calculus constituted one group of number-centric mathematical subjects seamlessly connected with each other. Second, there is a keen and continuing appreciation of the ability of algebra to make things clear and intelligible. Two phrases that are frequently used in the *Siromani* are *spasti-karana* and *sphuti-karana* (Colebrook 1817). The former literally means 'making clear', and refers to the process of describing specific operations through unambiguous instructions and directions. *Sphuta* is a more forceful word and is related to an event that occurs suddenly – from the rapid blooming of a flower to an instantaneous explosion (*sphota*). So *sphuti-karana* is the process of making something strikingly or brilliantly clear. (The words are also used to indicate corrections that improve the precision of a calculation.) The emphasis is therefore on the *intelligibility* of the algebraic process, and a literal translation into English of these Sanskrit words would be, accurately but awkwardly, 'intelligibilization'. The key behind this process is basically the idea of the algorithm. Bhaskara stresses the helpfulness of the introduction of unknown variables (*vividha-varna*, literally multi-coloured, but referring here to the use of colour names as already discussed for unknown variables). Bhaskara also talks about skill in handling unknown variables (*avyakta-yukti*), and the joy that his algorithms will bring to those who make computations (*ganak-ananda-karaka*).

The third characteristic is the frequent presentation of 'demonstrations' (*upa-patti*) – algebraic 'proofs' showing how adherence to and application of the accepted rules set out in various chapters of the *Bija-ganita* lead to the solution of various problems.

Much Sanskrit writing on mathematical-astronomical subjects is in terse verse. Some interesting material is therefore often not mentioned in the texts themselves, but there is ample evidence that in the instruction of the gurus to their disciples such material occupied a considerable fraction of interaction time. For example Jyesthadeva's text *Yukti-bhasa* (Sarma et al. 2008), which is like notes taken by a pupil during lectures given by the great Nilakantha of the Kerala school, was written down in Malayalam. This gives us a valuable opportunity to examine the way that mathematical materials were actually presented to young students by an experienced teacher. In particular, the derivation of the surface area of a sphere goes through an argument which involves cutting it into (bevelled) discs parallel to an equatorial plane, calculating the contributions to the area of the sphere made by each disc, and then taking the limit as the thickness of the disc goes to zero and the number of discs goes to infinity. In other words the argument is very similar to what would be used in a class room today when teaching calculus. This is something which we would not be able to understand if we were going entirely by Nilakantha's Sanskrit text, the *Tantra-samgraha*.

The point that is being made is that well before 1600, Indian mathematicians had a concept of mathematics of which algebra, equations, computation and calculus were integral parts. Each was connected with the others, and the methods used for deriving them often contained arguments that are familiar today. Nevertheless they differed fundamentally from European mathematics in their approach to the subject: this approach was number-centric rather than axiom-proof or geometry-centric.

Computational Positivism

One consequence of the computational revolution that took place in India after the introduction of the numeral system was the development of an attitude that may be called computational positivism (Narasimha 2003). It places computation and observation at the forefront. Although some hypotheses can be inferred as being at the back of the algorithms proposed in Indic *siddhantas, a priori* physical models, as well as a process of deduction based on axioms corresponding to the model or assumptions adopted, do not seem to play a fundamental role in Indic astronomy. The 'demonstrations' (*upapatti*) of the rationale for a procedure or an algorithm in Indian texts or commentaries are to be distinguished from formal proofs based on a minimal set of 'self-evident' axioms and deductionist logic. Indic demonstrations are best seen as constituting a different proving or validating culture, of which the chief elements were observation (aided by instruments when necessary), computation and verification (in the more exact sciences), and inference, combined and tied together by the force of the central concept of yukti, skill in general, but more particularly in reasoning and inference (Narasimha 2012).

In fact, the main goal of Indian astronomical schools was the achievement of *drg.ganit-aikya* (or sometimes *drg-ganita-samya*), which literally means the identity (or equality) of the seen and the computed. The effort therefore was to find algorithms or computational procedures which made the best predictions as determined by comparison with observation. If over a period of time discrepancies developed between computation and observation, the algorithms had to be tuned, modified or revised to bring computation back into agreement with observation – i.e. to restore *drg.ganit-aikya* (Sarma 1977).

This attitude of computational positivism had actual practical implications for the predictive methods used. For example Indian astronomers used patched ellipses for orbits, and took pleasure in making computations simpler and easier. But the Greeks would probably have been shocked by the idea; patched ellipses are not consistent with perfect, symmetric, beautiful nature.

Because of the primacy awarded to agreement between computation and prediction, the algorithms were periodically updated or revised to sustain agreement with observation. This has been well-illustrated by Billard (1977), and although his work has been criticized because the historical dating of the algorithms based on a best-agreement criterion turned out not to be always reliable (Pingree 1980), there is no doubt that computational positivism kept Indian predictions very competitive till almost a century after Newton.

Let us now consider developments in Europe beginning around 1500.

2   Francis Bacon and Inferred Axioms

> Aphorism I: *Man . . . does and understands only as much as he has observed of the order of nature in fact or by inference; he does not know and cannot do more.*
>
> FRANCIS BACON 1620 *The New Organon*

> *pratyaksena-anumanena* (By observation and by inference)
>
> NILAKANTHA (~1500) *Jyotir-mimamsa*

The ideas of Francis Bacon (1561–1626) are particularly relevant to our discussion for a special reason. During the last two centuries Asia has been grappling with the challenge posed by new knowledge from the West; in the 16th century the situation was just the other way round: it was Europe that was trying to tackle the challenge of new Asian knowledge.

To see this, let us first recollect Bacon's fascination with the many products of the 'mechanical arts' as he called them (today we would call it technology). He saw products in use in the Europe of his time that were unknown to his ancients, namely the Greeks and Romans. In one famous passage in the *Novum Organum* he selected gun powder, the printing press and the compass as three inventions which had changed the world far more than 'any sect, empire or star'. At other places in his writings he expressed his astonishment at the technology behind the production of silk, and a variety of other substances that were then beginning to be familiar in Europe (e.g. sugar). To the best of my knowledge he never explicitly acknowledged that all these technologies and products came from Asia, particularly from China, India and the Islamic lands. Surely he would have been aware – as somebody who, during the four years preceding the publication of the *Novum Organum*, had held such important positions in King James's government as Privy Councillor, Lord Keeper and then Lord Chancellor – that silk came from China and sugar from India. The fact that they had not been invented in Europe was for him strong evidence that there was something fundamentally wrong in the philosophy that contemporary Europe was following or, as we might put it today, there was a serious epistemological problem with the then prevailing European knowledge system.

So Bacon sets out to devise a new philosophy or epistemology that would ensure that new products of the mechanical arts would be available for the improvement of the 'estate of man'. Indeed he says at various places in the *Novum Organum* that for him the ultimate goal was not the mechanical arts themselves but rather the construction of a philosophy that was *truer* (in a special sense that we shall discuss below) than the one that Europe had been following. As much of the latter had been inherited from the Greeks he was severely critical of them (Gaukroger 2001). He poured contempt on Aristotle's philosophy (dismissed as that of 'a quack'), Plato (who was accused of spinning 'idle webs of speculation') and the rest. His objective therefore was to replace the prevailing views of medieval Europe by new ones, and to argue for a different view of 'science and technology' (as we would call it today). For example in a letter to King James he talked about the incompetence of the syllogism, suggesting that Greek-style argument was totally ineffective. He went on to emphasize the need for a systematic approach to knowledge that had to be promoted with the active support of the state; in other words it was a national task, one for the kings. An important part of his philosophy was the promotion of observation and experiment. In order to arrive at the truth nature had to be pursued, hunted and even harassed, but she also had to be wooed and even married. He was convinced that there were laws of nature and that they were inexorable, so defying them was not an option. The whole idea was therefore to discover how to *master* nature by *obeying* her own laws.

Although Bacon objected to Greek axioms as too speculative, his proposal was not to abandon or reject axioms altogether. Instead he insisted that they had to be based on *inference from observation and experiment*. Now inference has always seemed to be a somewhat uncertain proposition, and Bacon admitted that *inferred* axioms could prove to be wrong. But he was optimistic that such errors could be discovered quickly, and then new axioms would have to be formulated and tried out; 'truth' would thus eventually be discovered. He firmly believed that truth emerges more quickly from error than from confusion.

These philosophical ideas must have seemed revolutionary in Europe. But it is interesting for a contemporary Indian scientist to look at Bacon's proposals and assess what Indic reactions to them might have been at the time. My Indic ancients would have applauded the primary role that Bacon assigned to observation and experimentation. Contrary to the popular stereotype Indian scientists and philosophers unanimously awarded first place to *pratyaksa*, observation and perception, and in many fields (in particular chemistry) were great and scrupulous experimenters. Similarly they would have enthusiastically agreed with Bacon's criticism of the speculative character of the axioms often formulated by the Greeks, and would have told him that they had always known that inference was the only appropriate method of reasoning for gaining the most acceptance-worthy (or technically reliable) knowledge (Narasimha 2012). They were keenly aware that their conclusions were in principle tentative – after all, Nilakantha had explicitly said that astronomical siddhantas had to be continually examined and revised (Sarma 1977).

But Indians would have been surprised by Bacon's continued commitment to the idea of axioms, and his suggestion that the ultimate goal was *inferred* axioms. This was in fact the true novelty in Bacon's proposal – a novelty that was a fusion of Greek and Eastern ideas about knowledge. Universal knowledge was contained in the *axioms*, but they could not be found by the speculative, whimsical methods of the Greeks; instead they had to be the outcome of observation/experiment and inference. These were incidentally the two *pramanas* of philosophy that all Indic schools accepted as primary; only some would add other *pramanas* to the list. But their objective, for Bacon, was achieving a more universalizing knowledge than Indic scientists would have demanded, or considered possible.

These ideas had an extraordinary influence on thinking in Britain, and immediately thereafter on the European continent as well. Indeed, we can say that science developed in Europe in coming centuries very much along the lines that Bacon advocated, namely a powerful combination of a desire to master nature for the benefit of mankind, a belief in her inexorable laws, and the fundamental weakness of Greek-style logic; nevertheless axioms – a new kind of axioms – were the ultimate goal, and they had to be inferred by observation and experiment, abandoning them if they failed till the correct ones were eventually discovered. This faith in the existence of 'correct' or 'true' axioms, and of man's ultimate ability to discover them, was once again an idea that, I believe, would have surprised Indians. So would Bacon's equation of knowledge with power, and truth with utility, although utility in itself was always considered a legitimate demand in Indic thinking.

During the following centuries the Baconian programme turned out to be extraordinarily successful, for it transformed Europe and, eventually, the whole world. It still continues to dominate current thinking in much of science and technology.

3   Descartes and Barbarous Algebra

The next great milestone in the rise of European science was Rene Descartes (1596–1650). He is often credited with having introduced algebra into European mathematics. This was a momentous step and one that, in retrospect, changed the very concept of what mathematics was in the West. One of the interesting things about Descartes' work is that he attempted to algebraise geometry. In his famous book *Discourse on Method* there is a kind of appendix titled *Geometry*, published in 1637 (Smith and Latham 1950). His objective in introducing arithmetical/algebraic terms into geometry was *greater clarity* – precisely the same as recognized much earlier in India by Bhaskara, as we saw above.

We have already seen the sources of this clarity: first, the introduction of a notation for objects or variables, known and unknown, and clear rules for their manipulation; second, the formulation of an algorithm, reducing the solution of the problem to a mechanical process involving a finite sequence of instructions. When geometry is replaced by algebra, the often seemingly convoluted logic that is characteristic of the methods of Euclid's geometry are replaced by the task of manipulating symbols according to arithematical operations. This preference becomes apparent already in India during the times of Aryabhata and Brahmagupta (5th to 7th century), before it culminated in the Bija-ganita of Bhaskara. Descartes realized that the algebra that he was using was not European in origin, for in his work *Rules for the Direction of the Mind* he called it a *barbarous* art: there was no doubt that the name of the subject was itself alien to Europe. However it looks as if Descartes was puzzled about the history of the subject. As Mahoney (1980) points out, Descartes asserted on the one hand that algebra was a reconstruction of mathematics that the Greeks had known but had withheld from later generations deliberately – meanly in fact. On the other, he also claimed that *he* had created a mathematical method that the Greeks had known nothing about. The truth clearly is that his algebra was something which was basically imported from elsewhere – barbarous as he called it, not only in name but also in content, and furthermore that algebra had already been present in Italy since the 13th century when it acquired the subject from the Arabs and going back to India. Descartes introduced algebra to *Western* Europe.

In that book on geometry Descartes explained that he was designating each line with its specific length by a single letter such as *a, b*, etc. rather than by symbols like AB and CD, as we see it even today in texts on geometry derived from Euclid. By a^2 (which he often wrote as *aa*) he still meant only simple lines, but he named them squares etc. The translators of *Geometry* (Smith and Latham 1950) point out that at the time Descartes wrote his book a^2 commonly meant the area of a square whose side was *a*, and b^3 meant the volume of a cube whose side was *b*; b^4 for example was considered unintelligible, for no geometry in higher dimensions was conceivable at the time. Algebra liberated the concept of higher powers in number from having to make a direct connection with higher dimensions in geometry. One cannot help wondering at this point why Indians had no problems with such higher powers. For example the great Indian mathematician Madhava (1349–1425) had already written down infinite series for trigonometric functions containing arbitrarily high powers of the argument well before Descartes (Sarma et al. 2008). There is no evidence that Indians saw fourth and higher powers, in fact arbitrarily high powers, as a mystery. This is not a surprise, because the decimal place-value system can be seen as elegant shorthand for what is actually a series in powers of 10 (thus $3141596 = 6 \times 10^0 + 9 \times 10^1 + 5 \times 10^2 + 1 \times 10^3 + 4 \times 10^4 + 1 \times 10^5 + 3 \times 10^6$). This is in fact one other specific instance of the sense in which number was a logical prior of geometry in India.

So algebra spread from its early Indian roots in the 7th century through al Khwarizmi and the Islamic lands to the Europe of Viete and Descartes over the centuries. A partial history of this connection has been provided by Rashed (1994).

4   Algebra in Europe

It is appropriate here to present the conclusions from the detailed and enlightening analysis of the origin and development of algebra in Europe made by
Mahoney (1980). He correctly points out that 'the most important and basic
achievement at the time was the transition from the geometric mode of
thought to the algebraic'. The algebraic mode of thought was characterized by
an operative symbolism which was used to describe mathematical relations
between different, *a priori* unknown variables. To the extent that algebra han-
dled symbols it could be seen as more abstract than a subject like geometry,
which concerned itself with visible objects such as straight lines, circles, triangles and so on. In spite of the work of Diophantus Greek mathematics was
generally of a totally different character and displayed no great facility in handling either number or symbols (Cajori 1928). Mahoney points out that Greek
mathematics was intuitional and dependent on physical ontology.

Mahoney goes on to describe how European universities taught Euclid to
prepare students for Ptolemaic astronomy, whereas algebra, counted as an art,
was the concern of merchants, not of scholars. A leading pedagogue in the
16th century, Ramus, considered algebra vulgar. However algebra eventually
gained acceptance, chiefly because of the clarity and insight that it provided.
The growth of algebraic notation was helped by the advent of the printing
press and the printed book. The British philosopher John Locke said it was possible that those who were ignorant of algebra could not imagine its wonders.
Mahoney quotes Lavoisier, in his *Nomenclature of Chemistry*, as saying 'Algebra
is the analytical method *par excellence*; it was invented to facilitate the labors of
the mind, to compress into a few lines what would take pages to discuss, and to
lead finally, in a more convenient, prompt and certain manner to the solution
of very complicated questions'. These echo the words of Bhaskara. There can
be no doubt that what happened in Europe was the algebraization of mathematics in the centuries following Descartes. Algebra made mathematics 'intelligible', and analysis began to mean algebra. 'Analytical geometry' revealed the
algebraic relations not only for the triangle and circle but also between such
geometrical figures as the conic sections (ellipse, parabola, hyperbola).

As part of this process of algebraization the idea of the algorithm was also
absorbed, and found an enthusiastic proponent in Leibnitz. He even saw the
rules governing derivatives (i.e. instantaneous rates of change) as constituting the *algorithms* of differential calculus – echoing again the very Indic view
described earlier.

The profound influence that the advent of algebra had on mathematics itself
in Europe can be measured in part by examining the contents of mathematics

books that were published in Britain, beginning in the very late 15th century. Using the digital full text data bases of earlier English books that are now available, a fascinating analysis has been carried out recently by Wardhaugh (2009). The analysis reveals, first of all, that the word mathematics itself, which had an average frequency of appearance of about 5% during the century from 1481 to 1580, quadrupled during the century 1691 to 1790. The references to geometry *declined* from a little more than 22% in 1481–90 to about 10% in 1791–1800. On the other hand arithmetic went up from 0 in 1521–30 to around 10% in 1791–1800; and algebra, which did not exist before 1520, remained very low (less than 1%) till about 1640, ended up at about 3–4% towards the end of the 18th century. It is remarkable that, like algebra, arithmetic was also conspicuous by its absence before 1520. If we think of the two subjects together as number-centric it is clear that over a period of nearly three centuries they emerged from nothing in the early 16th century to surpass the references to geometry by the end of the 18th century. Although these statistics have certain limitations that Wardhaugh discusses in some detail, there is no doubt about the decline of geometry and the growth of algebra and arithmetic, coinciding with a remarkable rise in *mathematics* during the 16th to the 18th centuries in Britain. This change establishes the redefinition of mathematics in Europe.

It is clear that many of the key ideas that marked the advent of algebra in Europe – handling unknown variables, making mathematics 'intelligible', introducing a symbology along with rules for manipulating symbols, formulating algorithms as a finite set of unambiguous instructions that could be carried out mechanically – these and other processes associated with them were already known and used in India several centuries earlier. The scientific revolution that occurred around the 17th century is often credited to the *mathematization* of science. Galileo made the famous statement that the book of nature is written in the language of mathematics, but for him the characters of that language were still triangles, circles and other geometrical figures. But triangles and circles were not new; they were very much a part of Greek astronomy, as exemplified by Ptolemy and others before him. The mathematization of parts of science occurred much earlier in different ways at different times in history in countries such as Babylonia, Greece, India, China and elsewhere. What was special to 17th century Europe was *not so much mathematization as algebraisation*, but of course this event changed the very concept of what mathematics was, and thence led to an explosive growth of mathematics. It is really this event that the quote from Hermann Weyl at the head of this essay highlights.

5   Isaac Newton and Inferred Axioms

Newton took the common and familiar things and made them universal.
H.W. TURNBULL 1947

Isaac Newton occupies a singular position in the rise of science in Europe. From the point of view espoused in this essay his greatest achievement lay in carrying out the Baconian programme of discovering 'inferred axioms'. What we today call Newton's laws of motion appeared, in Motte's translation (1848) of Newton's *Principia Mathematica* (from Latin to English), under the Chapter heading 'Axioms' – on the first page of the book, in the time-honoured Euclidist style. They were axioms in the sense that Newton devoted two books of his work (out of three) to the task of deducing the logical consequences of those axioms in a variety of situations. And the mathematics he used for doing this was directly Euclidist; geometrical constructions, lemmas, theorems, corollaries, proofs and QED mark the two books. Book III however is very different in character because it is not very Euclidist; instead it is *drawing inferences from observations* (as in classical Indic science); in addition to the QED ('*quod erat demonstrandum*' or 'what had to be demonstrated') there appears a QEI ('*quod erat inveniendum*' or 'what had to be inferred'). But Newton takes the further step of relating them to the theorems and propositions of Books I and II. In doing this Newton was mixing a Greek geometrical tradition with something that was then entirely new – something that Francis Bacon had identified as the path to follow. I think Newton's brilliant successes established the credibility of the new epistemology that Bacon had proposed in his works, and laid a solid philosophical foundation for the spectacular growth of science and technology in Europe and (later) in North America and elsewhere.

A curious problem about the *Principia* is that there is little algebra in it. This was clearly not because Newton did not know the subject, for he was familiar with the work of Descartes and had himself made important contributions to it (e.g. the binomial theorem; see Turnbull 1947). There has been much debate among scholars about why Newton continued to use classical Euclidist methods (instead of the algebra he knew and the calculus he re-invented) in the first two books of the *Principia*. There is some indication that he considered the geometrical method more 'dignified' – recall the characterization of algebra as barbarous by Descartes, and vulgar by Ramus. From one point of view the *Principia* represents in part the swansong of an orthodox geometrical tradition; for Newton's successors in Europe, culminating with Leonhard Euler, proceeded with a vigorous programme of algebraising Newton, and in the process completely transformed both the subject and its methods within a few decades of Newton's death. And the classical mechanics that we study today as deriving from Newton is in fact not at all in Newton's language: $F = ma$ is Euler, not Newton. When Chandrasekhar (1995) wrote his commentary on the *Principia* he intended it 'for the common reader'. It was clear to Chandrasekhar that the 'common reader' would find algebra and calculus easier to understand than the complex logical arguments that Newton himself used in deriving his results. When we compare Newton's original argument side by side with its modern algebraic version in Chandrasekhar's book, the difference between the two is brilliantly clear – bleak on one side, sunshine on the other.

Newton's Rules of Philosophical Reasoning

The epistemological switch we already noted in the *Principia* between Books I and II on the one hand and Book III on the other was fundamental not only to the brilliant success of Newton's approach, but to the rise of science in Europe, and attests to the remarkable insight of Francis Bacon that led to his proposal for a 'new Organon' – a new instrument of knowledge or, as an Indic scientist or philosopher might have called it, a new *pramana*, a European *navya-nyaya*, 'new logic'.

Newton's switch is mediated by a brief digression headed 'Rules of Philosophical Reasoning', sandwiched between Book II and Book III of the *Principia*. These rules are formulated as follows:

Rule I: We are to admit no more causes of natural things than such as are both true and sufficient to explain their appearances.

Rule II: Therefore to the same natural effects we must, as far as possible, assign the same causes.

Rule III: The qualities of bodies, which admit neither intension nor remission of degrees, and which are found to belong to all bodies within the reach of our experiments, are to be esteemed the universal qualities of all bodies whatsoever.

Rule IV: In experimental philosophy we are to look upon propositions collected by general induction from phenomena as accurately or very nearly true, notwithstanding any contrary hypotheses that may be imagined, till such time as other phenomena occur, by which they may either be made more accurate, or liable to exceptions.

The first is chiefly a Rule of parsimony. In his brief explanation following the statement of the Rule, Newton supplies an Occam type 'razor' of his own, saying 'more is in vain when less will serve'. This implies that nature does nothing in vain and is 'pleased with simplicity'; there can be no superfluous causes. From this view of nature follows the idea that nothing more than a sufficient explanation, i.e. the simplest possible, should be admitted or accepted.

Rule II is a consequence of Rule I and demands that, as far as possible, one should not play with multiple causes for the same (or similar) effect. Once the simplest cause is found, we should not offer other possible causes. This is a powerful rule, as seen from the examples that Newton gives. We must look for the same causes for respiration in man and animal. The light from the fire in the kitchen and that from the Sun must not be attributed to different causes.[3] Similarly the reflection of light on the Earth and on the planets.

Rule III takes the idea further, and is a demand for accepting universality, based on experiment and observation. If all bodies on earth are attracted to the earth (by gravitation), if the tides in the sea are the consequence of the gravitational attraction of the moon, if the planets attract each other and the comets are attracted to the Sun and so on, we must admit the universal principle of gravitation. He upholds the primacy of experiment over 'vain fictions'. And 'Nature ... is wont to be simple, and always consonant to itself', as Bacon said in different words. There was in fact no other way of knowing as far as Newton was concerned.

Finally there is Rule IV, which is about the value of induction. This is very close to Bacon's concept of 'inferred axioms': these may have to be abandoned and replaced by superior propositions if new evidence appears, in the shape of new phenomena, more accurate measurements etc. But the 'argument of induction may not be evaded by hypotheses'.

Looking at these rules and Newton's commentary on them, we can see Indic 'rhythms' in them in several places. 'More is in vain when less will serve', of Rule I, is rather like an ancient *Samkhya-sutra* due to Kapila, who rejected the introduction of an additional principle on the grounds that 'it is inappropriate to postulate [/conceive] [unnecessary] entities (Samkhya-sutra 5.30, Narasimha 2011). Experiment and observation as primary sources of knowledge (Rule II), and inferring from phenomena (Rule IV), were (as already pointed out) universally accepted as the leading pramanas by all systems of Indic philosophy (including Hindu and Buddhist), going back in history for more than 2000 years. But some of the rules would have seemed alien to Indic philosophers, in particular those concerning nature's (alleged) simplicity and universality. Chaos is widely prevalent in nature, wind and water move in apparently

---

3 Of course we now know that kitchen fire is the result of chemical reactions whereas the fire in the Sun is produced by nuclear reactions, so Newton's inference is incorrect in this example. Indic scientists would not have been surprised.

unpredictable ways. It is now known that even the planets, which generally follow regular orbits, are not without a chaotic component in their motion; the celestial clockwork is therefore not as regular as earlier thought. Indians would probably have argued that these phenomena are counter-examples to Newton's very deterministic axioms. They admired simplicity in algorithms, but did not extend it to universal physical laws! As Playfair (1790) noted the rules Indians devised for making astronomical predictions were ingenious, *simple*, remarkable and surprisingly accurate, but they had no 'theory'. Similarly Indic philosophers sought unity in the diversity of the universe. Nevertheless neither the very specific rules of the mathematical astronomers nor the general principles of the philosophers stood for the kind of *physical* simplicity and universality that Newton was seeking through inexorable laws of nature.

6   The Flowering of Algebra and Analysis in 19th c. Europe

The 19th century saw the growth of algebra and the related concept of equations to unsurpassed heights in Europe: barbarous algebra had been assimilated and transformed beyond recognition from its Indo-Arabic roots. The very fact that Euler's equation $F = ma$ is now so indelibly associated with Newton's name, and is remembered in preference to the law as Newton himself stated it in prose, provides compelling demonstration of the power and clarity that algebra brought to exact science. As this move towards algebraisation continued in the 19th century, one striking achievement was the algebraisation of logic that George Boole achieved in England. It is interesting that George and his sister Mary were familiar with and often discussed Indian ideas about logic and algebra (Ganeri 2001).

The rapid and extraordinary developments that took place in algebra, calculus and analysis formed a major part of the European scientific revolution that left Asian mathematics well behind by the end of the 18th century. Indeed as the knowledge of algebra spread across Europe it appears to have branched out in two directions. The first was the ability to handle in some rational way quantities that were unknown; that was certainly the beginning of algebra in India. There was however a second branch in Europe that distinguished it from the earlier developments in India. (For example: an nth order equation has n roots, whose truth demands that we introduce 'imaginary' numbers involving $\sqrt{-1}$.) This was the subject of analysis, a concept which goes back to the Greeks. Here algebra was seen as helping in talking about equations – once again nothing new to Indic mathematics – but also talking about the nature of the solutions of the equations without actually solving them. This 'art of analysis' which eventually developed to great heights in 18th century Europe, in particular in the hands of Euler, had a dramatic effect on the complexion of both mathematics and physics in the West. It marked a radical departure from the geometrical roots of western mathematics, and was highlighted by Lagrange's celebrated boast in his work *Mecanique Analytique* that his book had no diagrams: it was strictly symbols and equations and algebra. (By the way Indian texts in astronomy and mathematics are also singularly bereft of any diagrams, often leading to the charge that thereby they are rendered more difficult to understand.) This was the big shift that occurred in Europe from the *Principia* of Newton (published 1687) to Lagrange's book of almost exactly a hundred years later (1788).

In parallel with the development of algebra was that of calculus as well. In India the development of calculus was marked by episodes spread over several centuries, starting from the concept of an instantaneous rate of change (the derivative), through second order interpolation schemes for obtaining functional values at arbitrary intermediate values of the argument, to integration, infinite series and the idea of limits at both ends: the infinitesimal and the infinite. All of this was one single package, in Europe as in India. One idea, or one technique, seems to have led inexorably to others in both civilizations. The explosive growth of mechanics, following the idea of inferred axioms that Bacon proposed and the laws of motion that Newton stated, quickly led to an orgy of new equations, combining algebra and calculus, driven in particular by the needs of mechanics. Mechanics and physics were not as much a part of exact science as astronomy in India, so that while algebraic equations were quickly invented and adopted, they did not lead into a new language of quantitative physics in India. This of course is where social, cultural, geographical, economic and political factors play a great role.

An interesting development here was that, side by side with the analysis which attempted to derive knowledge of the nature of the solutions without actually solving the equation, the need for hard predictions demanded the development of computational techniques. In India this had happened *before* the development of calculus, and was driven by astronomy; computation had for long been an integral part of the package. We have already seen that Francis Bacon had recognized that the ability to make calculations with large numbers was important for the development of useful knowledge. Perhaps the first European scientist to pursue a vigorous and focused programme that awarded to computation a prominent position in mathematics and physics was Leibnitz. He realized that the concept of the algorithm could have an extraordinary impact in the way exact science was done. And so the algorithm now became an object of serious analysis and study in Europe. A finite number of serialized, clear and unambiguous instructions that can be followed mechanically became a powerful mathematical tool, although devising algorithms for this purpose required great ingenuity beyond the scope of mechanical calculation. This division has had a profound effect on the way that mathematics is done. Once one has a method of handling numbers such as the Indic system, reducing astronomy or mechanics to a set of algorithms almost becomes equivalent to having found the solution to the problem, and this is the way that India had followed. Following the advocacy of Leibnitz, computational methods also saw a period of extraordinary development in Europe.

As writing an equation became almost synonymous with making an intelligible and precise statement, more and more of the world began to find it easier to grasp what an equation said than to understand a convoluted assertion in prose. A new artificial language had come into being, and was growing in richness and complexity, going far beyond the simple equations of a thousand years earlier. By the 19th century exact science was dominated by equations as new ones were found to describe ever growing domains of physical knowledge; and this has continued to this day. To recount only a short list of the major landmarks here, by the middle of the 18th century Newton's laws had been converted to precise equations by Euler not only for particles, but also for rigid bodies, elastic solids and inviscid fluids. In the 19th century equations were found by Navier and Stokes for describing hydrodynamics; by Clerk Maxwell for describing electromagnetism; and by Ludwig Boltzmann for describing the motion of molecules in a gas. These were followed in the 20th century by the famous equations of Einstein, Schrodinger, Heisenberg and others. Equations became so central that Dirac (1963) said that his equations were smarter than himself. This is understandable because there are many equations that express a law unambiguously, but cannot still be solved in the classical sense of analysis. Put in other words there are equations that seem to contain truth but are so difficult to solve that their implications are not easy to work out. The Navier-Stokes equations of fluid flow provide a famous example: the laws governing the motions of air and water that we see all around us on our planet are known, but their consequences for describing even the simplest turbulent flows (as in mundane plumbing, for example) cannot yet be derived in a scientifically satisfactory way. Equations were even found to be capable of handling uncertainty in precise ways: Schrodinger's equation in quantum mechanics is an excellent example.

The arguments above suggest that in the history of astronomy and mathematics, there have been at least two ways of pursuing an exact science: one valuing axiom-based formal proof, the other observation-driven computation. Through processes of fusion and diffusion these two views led to the Baconian system of inferred axioms (i.e. inferred from observation and experiment), mixed with the algebraization and algorization of a previously largely geometric mathematics. This new system has held sway over global science for the last three centuries.

Bibliography

Billard, R. "Aryabhata and Indian astronomy," *Ind. J. Hist. Sci.* (1977) 12: 207–224.

Cajori, F. *A History of Mathematical Notations.* La Salle IL, USA: Open Court Pub. Co.

Chandrasekhar, S. (1995). *Newton's Principia for the Common Reader.* Oxford: Clarendon Press.

Colebrook, H.T. (1817). *Algebra, with Arithmetic and Mensuration, from the Sanskrit of Brahmagupta and Bhaskara.* Delhi: Sharada Pub. House. Reprint 2005.

Datta, B. and A.N. Singh (1935). *History of Hindu Mathematics.* Bombay: Asia Publishing.

Dirac, P.A.M. "The Evolution of the Physicist's Picture of Nature," *Scientific American* (1963) 208: 45–53.

Ganeri, J. (2001). *Indian Logic: A Reader.* UK: Curzon Press.

Gaukroger, S. (2001). *Francis Bacon and the Transformation of Early Modern Philosophy.* Cambridge: Cambridge University Press.

Hayashi, T. (1995). *The Bakhshaali Manuscript.* Groningen: Egbert Forsten.

Ifrah, G. (1998). *The Universal History of Numbers.* London: Harvill Press.

Jardine, L. and M. Silverthorne (2000). *Francis Bacon: The New Organon.* Cambridge: Cambridge University Press.

Mahoney, M.S. (1980). "The beginnings of algebraic thought in the seventeenth century," chapter 5 in S. Gaukroger (ed.), *Descartes: Philosophy, Mathematics and Physics.* Barnes and Noble Books. Also www.princeton.edu/~hos/Mahoney.

Minkowski, C. (2004). "Competing cosmologies in early modern Indian astronomy," in C. Burnett, J. Hogendijk, & K. Plofker (eds.), *Ketuprakasa: Studies in the history of the exact sciences in honor of David Pingree* Leiden: Brill, pp. 349–385.

Narasimha, R. "Sines in terse verse," *Nature* (2001) 414: 851.

———— "The Indian half of Needham's question," *Interdisciplinary Science Reviews* (2003) 28(1): 54–66.

———— "Axiomatism and Computational Positivism: Two Mathematical Cultures in the Pursuit of the Exact Sciences," *The Economic and Political Weekly* (2003) 38: 3650–56.

———— "Epistemology and Language in Indian Astronomy and Mathematics," *J Indian Philos* (2007) 35: 521–541.

———— (2008). "Chequered histories of epistemology and science," in Bharati Ray (ed.) *Different Histories.* Chapter 6, pp 89–122. (PHISPC. Vol.14 pt.4). Delhi: Pearson Longman.

——— (2011). "Nature views culture," in R. Narasimha and S. Menon (eds.) *Nature and Culture*, pp. 321–339 (PHIPSC Vol.14 pt.1). New Delhi: Centre for Studies in Civilizations. Munshiram Manoharlal.

——— (2012). "Pramanas, proofs, and the yukti of classical Indic science," chapter 5, pp. 93–109 in Bala A. (ed.), *Asia, Europe and the Emergence of Modern Science.* New York: Palgrave Macmillan.

——— (2013). "The yukti of classical Indian astronomy," in B.V. Sreekantan (ed.), *The Foundations of Science*, pp. 277–294 New Delhi: Centre for Studies in Civilizations Pearson.

Newton, Isaac. *The Principia.* Translated by A. Motte, A. (1848). New York, NY: Prometheus Books (Reprint 1995).

Pingree, D. "Response," *J. Historical Astronomy* (1980) 11: 63.

Playfair, J. (1790). "Remarks on the Astronomy of the Brahmins," *Transactions of the Royal Society, Edinburgh* 2(1): 135–192.

——— (1798). "On the Trigonometry of the Brahmins," *Transactions of the Royal Society of Edinburgh.* In *The Works of John Playfair*, Esq. 3, 255–276.

Ramasubramanian, K. and M.S. Sriram (2011). *Tantrasangraha of Nilakantha Somayaji.* New Delhi: Hindustan Book Agency.

Rashed, R. (1994). *The Development of Arabic Mathematics.* Kluwer.

Sarma, K.V. (1977). *Jyotirmimamsa of Nilakantha Somayaji.* (Panjab University Indological Series 11), Hoshiarpur, India: Vishveshvaranand Vishva Bandhu Institute of Sanskrit and Indologial Studies.

Sarma, K.V., K. Ramasubramanian, M.D. Srinivas and M.S. Sriram (2008). *Ganita-Yukti-Bhasa of Jyesthadeva.* New Delhi: Hindustan Book Agency.

Shukla, K.S. and K.V. Sarma (1976). *Aryabhatiya of Aryabhata.* New Delhi: Ind. Natl. Sci. Acad.

Smith, D.E. and M.L. Latham (1950). *The Geometry of Rene Descartes.* New York: Dover.

Turnbull, H.W. (1947). *Newton: The Algebraist and Geometer.* In: *The Royal Society Newton Tercentenary Celebrations*, pp. 62–72.. Cambridge: Cambridge University Press.

Wardhaugh, B. (2009). 'Mathematics in English printed books, 1473–1800: a bibliometric analysis'. *Notes & Records, Royal Society* 63: 325–338.

Weyl, H. (1928). *The Theory of Groups and Quantum Mechanics* (1931 edition trans. H.P. Robertson, 1931). New York, NY: Dover.

CHAPTER 8

The Transfer of Geographic Knowledge of Afro-Eurasia in the "Bright" Middle Ages: Cases of Late Medieval European Maps of the World

Hyunhee Park

1   Introduction

It is well known that Henry the Navigator and Christopher Columbus read Marco Polo's travel account, with its fascinating stories about the world of eastern Eurasia, and that it inspired them to explore these lands personally. (Skelton 1958: 16–17) Polo's account significantly influenced medieval European knowledge of the world. It was one of the first accounts of Asia that provided Europeans with a lot of new and reliable information; because of this, it became popular, too. People increasingly utilized the traveler's account to write geographic works during the fourteenth and fifteenth centuries. Polo's impact has created the misconception that European knowledge about the world began with him. In fact, however, geographic knowledge of the wider world continuously traveled to Europe from the East both before and after Marco Polo, through a variety of middlemen, books and maps, thanks especially to the Crusades and the Mongol empire connections.[1] That is, European accumulation of new geographic knowledge about the world increased gradually and variously through the new political channels that emerged during the High and Late Middle Ages (1000–1500). Polo's book was significant, but only as a watershed in a longer intellectual trend. During the fifteenth century, Portuguese collected and examined a large quantity of this new geographic information, and used it to launch expeditions of global exploration that led to the first European navigation to Asia by rounding the horn of Africa. (Safier and Santos 2007: 461–468) The Spanish who chose to reach Asia traveling in the other, westerly, direction soon followed them.

Despite obvious clues to the significant role that the transfer of geographic information from Asia to Europe played in the transformation of European worldviews and the growth of Europe's global connections during the early

---

[1]   On the influence of Marco Polo's account or other contemporaneous accounts on European knowledge of Asia, see Jackson (2005), pp. 329–357.

modern period, little work has been done to examine the overall flows of this knowledge in Eurasian geography and cartography. A rich collection of cartographic materials is available to anyone who wishes to explore the level of geographic knowledge about the world that different Eurasian societies possessed. However, most current studies focus on individual materials that developed in a particular society or compare sources at a survey level.[2] The task remains, then, to compare and connect these materials in order to develop a clearer understanding of the broader trends of geographic and cartographic knowledge transfer from Asia into Europe and how that flow of knowledge gradually changed European understanding of the world at a crucial transitional period in global history.[3]

This essay will explore the transfer of geographic knowledge from Asia to Europe by focusing on two new types of maps – different from the traditional "T-in-the-O" form of map – that appeared in late medieval Europe and created breakthroughs in the depiction of the wider world including Asia. As an example of the first type, a world map submitted by a Venetian politician to the Pope renders the Afro-Eurasian continents with more accurate contours than previous maps and shows the Indian Ocean extending to the Far East. Maps of this type suggest access to Islamic cartographic knowledge of the Indian Ocean and Asia. These maps were soon followed by a second, more sophisticated type of map, exemplified by works such as Fra Mauro's map, that adopted the new coastlines introduced earlier by the first type of map, and which added details about the wider world including Asia based largely, but not exclusively, on information derived from Marco Polo's account. The revolutionary changes in both the style and content of these world maps testify to the changes in geographic understanding of the world that took place among some European intellectuals during late medieval times. Geographic data and concepts that Europeans gained through their increased access to Asian geography contributed to this breakthrough.

---

2 The most recent and comprehensive project about the history of cartography in world history done so far – and still underway – is the set of several volumes edited by J.B. Harley and David Woodward published by Chicago University Press. J.B. Harley, and David Woodward, eds., *The History of Cartography* (Chicago: University of Chicago Press, 1987–), 6 vols. There are many other significant volumes that focus on specific topics, some of which are cited in this paper.

3 Victoria Morse suggests that, while the discovery of new maps and new texts prompts us to reevaluate the cartography of the high Middle Ages, we should explore more fully changes that occurred in the fourteenth and fifteenth centuries. Morse (2007), p. 26.

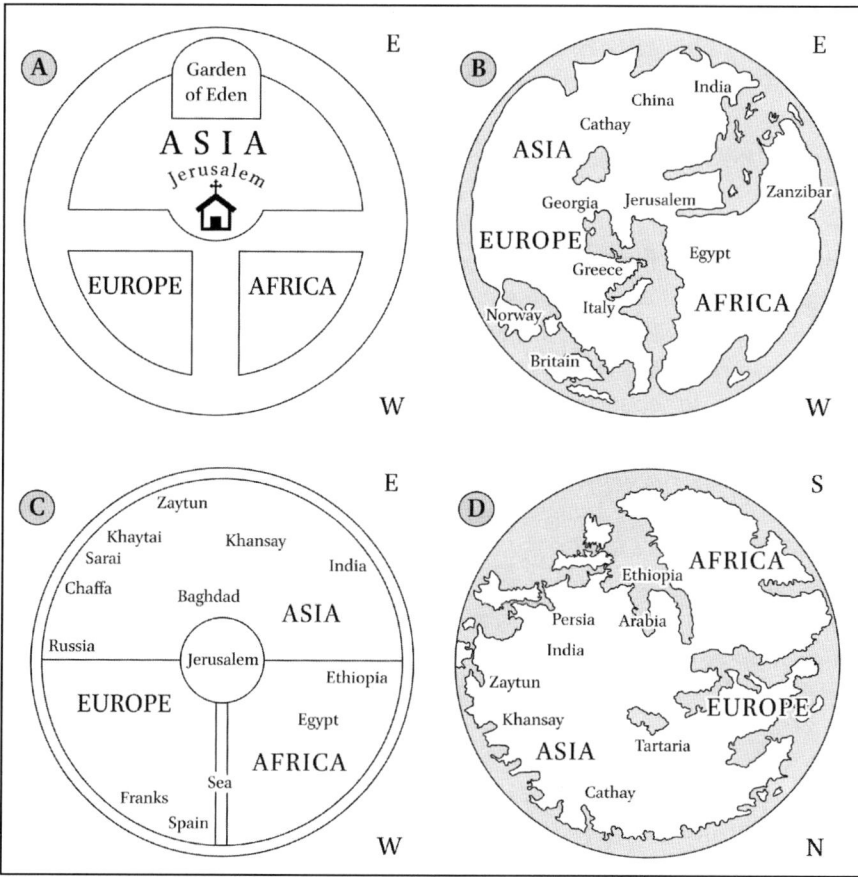

FIGURE 8.1 *Comparison of the outlines and contents of the four world maps discussed in this chapter. Reconstruction by the author. Map A: T-O Map of Medieval Europe. Map B: Sanudo-Vesconti world map. Map C: Armenian T-O Map. Map D: Fra Mauro Map.*

2   T-O Maps: The Most Dominant World Map Form in Medieval Europe

The "T-in-the-O" map (abbreviated as "T-O map" below) was the most dominant form of world map (called *mappa mundi*) that circulated throughout medieval Europe (see map A in Figure 8.1).[4] This form of map began to appear

---

4   There were other types of contemporaneous world maps, including the zonal maps that circulated at the same time. However, most of them are similar in their limited geographic

in the eighth century; as it diffused throughout the continent, it developed local variations in detail while it kept the same basic structure of land division. This structure separates the world's landmass into the three major continents of Asia, Africa, and Europe; at the center of the world lay Jerusalem, the holiest city in Christendom. These continents are placed close to each other, separated only by watercourses that flow west from the Garden of Eden at the top portion of the map, where T-O mapmakers placed the eastern end of the world. While the Hereford map of the early fourteenth century depicts the continents and watercourses in a more realistic way than previous maps, the shape of continental coastlines generally resemble the simpler forms found in T-O maps like the one described by Isidore in the seventh century. In these maps, Asia occupies the largest portion of space, almost half the map. However, the topographical information in these T-O maps is vague. For example, none of these maps bear any place names that designate China, one of the most prominent societies in premodern Asia. In place of China, T-O maps identify the eastern end of Asia as the site for the Garden of Eden, the place where Adam and Eve, the world's first people according to Judeo-Christian religious tradition, first lived. This understanding of Asia as a paradise detached from the real world repeats itself in contemporaneous geographic texts. (Akbari 2008)

How the T-O map of the world suddenly appeared in medieval Europe has mystified many historians of cartography because the map significantly differs from the realist cartographic tradition of Afro-Eurasia that the ancient Greeks and Romans developed between the fifth century BCE and the second century CE. (Harvey 1991: 9) Strabo and Ptolemy's longitudinal and latitudinal tables of Eurasia and North Africa show that ancient Greek and Roman geographers possessed a fairly accurate understanding of these broad regions. Greco-Roman advancements in mathematics and astronomy facilitated this remarkable achievement because it allowed them to locate places on the meridian based on empirical geographical measurement of the globe. This included the approximate locations of remote regions brought by travelers and merchants who had reached Central Asia and the Indian Ocean world regions. (See Aujac, et al. 1987) In the longitudinal and latitudinal tables of his *Geography*, Ptolemy of Alexandria placed Serica (meaning a country from which silks come) at the upper eastern part of the Eurasian continent, and another country called Sinae at its southeastern edge, within a space that is roughly equivalent to China. Ptolemy's *Geography* also locates the Indian Ocean; although he describes it incorrectly as enclosed, he correctly shows

understanding of the world. For more detailed discussions, see Edson (2007), Harvey (2006) and Woodward (1987).

that the body of water continues east to Sinae. The famous first-century Greek text *Periplus of the Erythraean Sea* demonstrates that the ancient Greeks had practical navigational knowledge about the Indian Ocean; as a merchant hand-book, it locates Thina at the eastern end of the Indian Ocean, and describes it as a place where merchants obtained silks.[5]

This type of realistic information about the wider world that Greek and Roman geographers developed, however, did not transmit to Europeans over the course of the tumultuous fall of the Roman empire that began in the fifth century with the Germanic migrations. The works of Strabo and Ptolemy were handed down to medieval Europeans only through short encyclopedic pieces, because their rich originals were lost. Long before these ancient legacies were recovered in the fifteenth century from Byzantine archives, however, they had passed along into the Islamic empires that arose as a new political and cul-tural force in the seventh century. Geographers in the Islamic world, along with those in other scientific fields, actively adopted Greek and Roman knowl-edge of world geography from the Byzantine populations that Muslim armies conquered; moreover, they improved earlier geographic knowledge by using updated methods to re-measure the longitudes and latitudes of major cities and incorporating new information that Muslim merchants and travelers con-tinuously brought back with them from their travels. Over time, generations of scholars and schools of Islamic geography produced a variety of world maps that plausibly described the major continents and seas of Afro-Eurasia. This continued for several centuries throughout the medieval period.[6]

In contrast, even after a new political order emerged in Europe, connec-tions with the wider world weakened, and European knowledge of the world diminished. Many of Europe's major societies no longer connected to the wider Indian Ocean trade networks that flourished across wider Eurasia, and therefore lost channels through which new information from the outside world could have come. At the same time, the rise of Christianity as a religious and political power began to dominate the worldview of Europeans. Therefore, it is not strange to see the gradual spread of T-O maps, facilitated by church fathers, that replaced the world's eastern lands with the Judeo-Christian para-dise. (Woodward 1987: 299) Despite variation in expression across place and time, the basic form of the T-O map survived, and the map continued to flour-ish throughout Europe and dominate its geographic conception of the world throughout the medieval period.

---

5   *The Periplus*, translated by Casson (1989).
6   Tibbetts (1992a & 1992b); Park (2012), pp. 56–90.

3   New European Circular World Maps of the Early Fourteenth
        Century in the Context of the Crusades and Its Islamic Connections

In the fourteenth century, there suddenly appeared world maps that differed considerably from the T-O maps. In one example, a revolutionary type of world map divided the major continents of the Afro-Eurasian world by the Mediterranean Sea and Indian Ocean and delineated the contours of the continents more closely to reality than conventional T-O maps had done. These new contours of the Afro-Eurasian landmass influenced many world maps of the late medieval period. One of the earliest surviving maps of this kind is the world map that Marino Sanudo the Elder submitted in 1321 to Pope John XXII (see map B in Figure 2), along with several sea charts of the Mediterranean regions and a political treatise that urged a new crusade against the Islamic world.

   Sanudo was a politician who was raised and educated in one of the famous Venetian families of the medieval period. As a sincere Christian, he presented himself as an active supporter of the Crusades, the wars that Christian Europeans waged against the Muslim kingdoms of the Middle East, in order to take their holy city, Jerusalem, from the Muslims who controlled it. This conflict, which lasted from 1095 to 1291, affected Europeans more than Muslims because it provided them with the opportunity to encounter people from the east, to look towards the eastern world, and to gradually end their long period of isolation from the wider Asian networks of travel and trade. Sanudo's maps, including a world map, and his treatise collectively demonstrate this. In addition to religion, then, men like Sanudo probably also wished to protect Venetian commercial interests in the maritime trade that had grown during this period.

   Sanudo's treatise, *Liber Secretorum Fidelium Crucis* [*The Book of Secrets of the Faithful of the Cross*], offers many strategies to win the European war against the Islamic world. This includes the cessation of all Christian trade with Muslim countries across the Mediterranean and the establishment of a Christian fleet in the Indian Ocean, which would seek to dominate and even subjugate its seacoasts and islands that fell along its principal trade routes.[7] His written sketch of the trade routes that crossed Persia and Egypt, and his description of the course of Indian trade from Coromandel and Gujarat, to Hormuz and the Persian Gulf, and to Aden and the Nile, is supported by his world map, which roughly outlines the East African coast, Arabian Peninsula, and Asian continent above the Indian Ocean. In other words, people could now understand the physical geographical shapes of the continent to the east by

---

7   Sanudo, *Marino Sanudo Torsello* (2011), pp. 49–66.

comparing the treatise with a world map, something the traditional T-O maps hardly allowed. Sanudo himself also asserts the importance of understanding the actual physical geography of the world in order to achieve a political goal, and he made many copies of his treatises and maps, which circulated widely. (Edson 2007: 62)

Another new feature that distinguishes Sanudo's map from the earlier T-O map is the use of compass rhumb lines, a feature often found in contemporaneous portolan charts that suddenly began to appear in the late thirteenth century. The portolan chart represents a new type of map that focuses on recording the actual navigational lines of Mediterranean regions. Because the world map submitted by Sanudo had been accompanied by several sea charts of this type, and these charts are identical to those belonging to the famous Genoese cartographer Pietro Vesconte (fl. 1310–1330), scholars assume that Vesconte drew Sanudo's maps, including the world map.[8] The discovery of a copy of Sanudo's treatise with Vesconte's signature, dated 1320, provides evidence of connections between Sanudo's maps and Vesconte's workshop. Still, one wonders how Vesconte received the basic cartographic information necessary to draw the world map in a realistic shape using the compass rhumb lines. Cartographers of portolan charts drew the compass rhumb lines on the sea charts for different parts of the Mediterranean Sea based on the actual navigation there. Some historians of cartography argue that the creators of the Sanudo world map based their construction on the traditions and methods of portolan charts. However, Venetian sailors, the most likely source of this information, were not yet fully active in the Indian Ocean, at least enough to understand the navigation of the Indian Ocean themselves. How did they draw other places in the wider world, like the vast Indian Ocean, where they had little navigational experience, using the same method? This question leads us to suggest that influences and information came from outside Europe.[9] That

---

8   Because the world map and other regional charts were drawn professionally, it would be a mistake to consider Sanudo the maps' author. Sanudo does not mention the name of the map's actual cartographer; however, some scholars found that this map was drawn in the same cartographic style as the maps that were made by a contemporary, the famous Genoese cartographer Pietro Vesconte (fl. 1310–1330), and therefore they believe that Vesconte must be the cartographer who made the world map that Sanudo submitted to the Pope. (Edson 2007: 63). Nowadays, some scholars call the map the "Sanudo-Vesconte" map, or simply the "Vesconte map."

9   Although European sailors gradually expanded the scope of their navigation over the centuries, and developed portolan charts for use in practical navigation, their knowledge had not yet advanced to the point that they could have provided the map's information about the Indian Ocean and upper part of Asia.

is, the fact that the scope of all surviving portolan charts are limited to the Mediterranean Sea – and therefore contain no information about the wider world – supports this view. In fact, another contemporaneous world map contained in a manuscript of Brunetto Latini's (c. 1220–1294) encyclopedic work, *Li Livres dou Tresor* [*Treasure House of Knowledge*], which resembles the Sanudo map and displays realistic contours of the world including the Indian Ocean, does not contain rhumb lines.[10] We can assume, therefore, that this kind of cartographic idea spread to some places in Europe during the fourteenth century. Either Sanudo or Vesconte acquired the same model to use as a base map for drawing a new type of world map that included new Afro-Eurasian contours. Moreover, the cartographer of the Sanudo world map also incorporated the elements of portolan charts that were found in his other maps.[11]

Where, then, did the new model for drawing Afro-Eurasian contours originally come from? Some scholars have already credibly suggested that Islamic maps, such as those drawn by the famous Muslim geographer al-Idrīsī, may have influenced Afro-Eurasia's new shape.[12] Al-Idrīsī was a Muslim scholar who worked in the court of Roger II of Sicily. The king sponsored al-Idrīsī's endeavor to collect available geographic knowledge of the world and compile them in a geographic treatise and set of maps. In the middle of the twelfth century, this project represented the greatest advancement in geographic knowledge of the world. Indeed the Sanudo world map looks very similar to the circular world map of al-Idrīsī, which circulated widely within the Islamic world at the time. Of course, discrepancies between the two maps exist, such as in their transliterations of place names, as Evelyn Edson argues. However, Piero Falchetta assumes that the Sanudo world map's cartographer simply copied the shapes of the Islamic world map and did not incorporate its place names because he was not able to read the Arabic words. (Falchetta 2006: 34) It is also possible that the cartographer of the Sanudo world map used an Islamic world map other than that of al-Idrīsī, considering that al-Idrīsī's map also drew heavily from earlier Islamic maps (such as those associated with Caliph

---

10   We do now know whether the world map found in Brunetto Latini's manuscript bore any relationship to the Sanudo world map. (Edson 2007: 69–70)

11   Another contemporaneous world map of a similar type is the circular map of Fra Paulino Minorita, which he included in his treatise, the *Demapa mundi* (ca. 1320). (Larner 1999: 134). Sanudo and Paulino probably exchanged ideas throughout their period of cooperation, sponsored by the Pope. (Edson 2007: 70)

12   See Edson's discussion of Joachim Lelewel and Tadeusz Lewicki as the first people to allege that Sanudo had seen an Arabic map, perhaps even one by al-Idrīsī himself. (Edson 2007: 68–69)

al-Ma'mūn and the Balkhī school). Many of these maps, including those that were developed after al-Idrīsī's map appeared, share a similarity in their realistic depiction of the Afro-Eurasian landmass. Some cartographers in Venice or Genoa, including the cartographer of the Sanudo world map, were probably influenced by standard Islamic world maps, which they had encountered through commercial or military contact. Similarly, Brunetto Latini, Sanudo's contemporary and a notable scholar and politician from Florence, also probably acquired a world map influenced by Islamic geographic works during his broad travels between Florence, France, and Spain – particularly Spain, which was profoundly influenced by Islamic science and culture – and included it in his encyclopedic work. (Sezgin 2005: 223) The influence behind Sanudo and Brunetto Latini, therefore, could be a type of Islamic world map bearing similar contours that includes the Mediterranean Sea and Indian Ocean. It is unconvincing to insist that European mapmakers all of the sudden created this wholly new kind of map based solely on European traditions. Thus the most plausible explanation for the creation of the Sanudo world map argues that its cartographer had access to an Islamic circular map or to a copy made by other European makers of similar world maps. The cartographer, then, combined the fairly accurate contours of the Afro-Eurasian continents with new popular European cartographic methods, such as the compass rhumb lines of the portolan chart, in order to produce the most up-to-date world map of its time for practical use.

Historically, many new types of maps were made based on the combination of old and new methods into new hybrid forms. Mapmakers sometimes revealed what kind of geographic materials they used to compile new types of maps, but sometimes they did not.[13] Sanudo did not reveal the cartographer behind the maps that he submitted to the Pope. Later scholars worked hard to argue that it was Vesconte. It is even more difficult to trace the Islamic sources that the cartographer used to draw this new type of world map. Nevertheless, we can assume the potential reasons why Sanudo may have withheld his map's sources if we explore the historical context of the map's creation. Sanudo himself had strongly urged against any contact with Muslims because they were the political enemy and trading rival of European Christians; at the same time, he knew that the source of his world map came from the Islamic world. Perhaps he did not want to reveal that knowledge. Such a historical contextualization

---

13   For examples that explicitly show how mapmakers combined different maps to draw a new map, see the productions of world maps made in the fourteenth and fifteenth centuries in China and Korea in Park (2012), pp. 100–107.

suggests that the European cartographers who first made the new, more realistic kind of world maps used Islamic sources without citing them.

After these maps appeared in the thirteenth century, many subsequent maps of various types began to assimilate the same realistic contour. Although we do not have direct evidence of the connections that existed between the first maps that bore the new shape of Afro-Eurasia and those that appeared in later periods, we can nonetheless assume that the first new cartographic style gradually spread and began to influence cartographic conventions during the late medieval period. This serves as credible evidence that the revolutionary change in Europe's world geographic understanding developed from outside connections.

While the fourteenth-century Sanudo world map demonstrates the innovation of physical contours in European maps of the world, it also contains reliable details, another new feature that makes world maps more realistic than before. The Sanudo map is also one of the first European maps to replace the Garden of Eden with China and Cathay (northern China) in eastern Asia. In addition to the more detailed information about Asia that Sanudo provides in his written treatise, the new place name, Cathay (Khātāi, the name that Muslim geographers used since the twelfth century to refer to northern China) derived from information that Europeans obtained from Prince Hayton of Little Armenia, a vassal state of the Mongols since the 1240s, through the Latin and French translations that they produced of his account about the realms of Asia. (Hayton 1989) More realistic information about Asia, newly acquired through Mongol connections, gradually proliferated in maps drawn during the thirteenth and fourteenth centuries, most prominently through Marco Polo's account of his travels east to places like China. The next section will discuss these later maps of newer type in greater detail, focusing on Fra Mauro's world map, in order to explore the influence of these connections on the accrual of new detailed information about Asia.

4   Adding New Detailed Information about Asia to European World
        Maps through Mongol Connections, and Travelers' Accounts

In the thirteenth century, while the Crusades were underway in the western part of Eurasia, a big political change occurred in the east: the rise of the Mongols from the northern steppes of Mongolia, and their conquest of most of the sedentary societies inhabiting Asia and Eastern Europe. They created the largest contiguous empire in world history, and thanks to this, facilitated a new level of long-distance contacts between the different societies of Eurasia.[14] It connected societies across Asia, such as China and Iran, more closely than had been the case during earlier periods and, more importantly to the history of premodern contact in Eurasia, linked Europe to existing Asian networks. Although the western part of Europe had been saved from the Mongol conquest, the fear that developed as the Mongols swiftly conquered vast portions of eastern Europe led political leaders in western Europe to dispatch envoys to the Mongol empire. Some of the envoys, like John of Plano Carpini (1182–1252) and William of Rubruck (c. 1220–1293), left first-hand accounts of the Mongols based on their travels to Mongolia.[15] In fact, some new detailed pieces of information about Asia that Sanudo incorporated into his written treatise came from the works of Carpini.[16] Besides, many European merchants acquired opportunities to travel to eastern Asia using the continent-wide networks that the Mongols had opened. The most famous traveler to claim that he had been to China was the Venetian merchant Marco Polo, whose travel account became one of medieval Europe's bestselling books and exerted a lasting influence on the European geographic understanding of Asia. In fact, other Europeans like Odoric of Pordenone (1286–1331) also claimed to have traveled to China and left behind their travel accounts, too. Thanks to them, we can find more concrete traces of Europeans on Chinese soil, as some archeological sources demonstrate. (Vogel 2012: 351–353) In short, regardless of the debates about the reliability of some of these travels, most notably Polo, there can be no doubt that many people visited societies across Eurasia – at least those that were affected by the empire's better developed transportation system – and, in the process, exchanged geographic knowledge across political and cultural lines on an unprecedented scale. Sometime after these historical changes occurred, new varieties of world-map types appeared, ones that incorporated new, accurate, and concrete information drawn from reliable travel accounts. Gradually, these new kinds of maps appeared in some cities throughout Europe.

---

14   Many studies have been done on the Mongol empire, yet scholars have only recently begun to reassess its influence and have called for more studies on various topics using sources in different languages showing cross-cultural interactions across the empire and beyond. For succinct reviews of studies of the Mongol empire, see Jackson (2000) and Morgan (2007), pp. 181–227.

15   On the accounts of the Christian missionaries and other to Mongolia and China during the Mongol period, see Jackson (2000), pp. 135–153 & pp. 256–289.

16   Ibid., pp. 26–27, footnote 27 in p. 352.

After the fourteenth century this new geographical information about the world led to greater innovations in T-O maps, which continued to circulate in Europe after the appearance of more realistic types of maps like the Sanudo map. And yet, the old T-O map form evolved into new forms that differed from earlier forms. For example, some cartographers began to incorporate more realistic depictions of the Mediterranean Sea and Indian Ocean in place of the long-used and narrow watercourses.[17] Some went in the opposite direction, however, providing new information while maintaining the T-O map's standard continental shapes, including the watercourses that flow from the eastern end of the world. For example, an Armenian T-O map drawn sometime after the thirteenth century labels China as "Khātāi (Cathay)" (see map C in Figure 8.1).[18] This makes the Armenian map (along with the Sanudo world map) among the first extant medieval European maps to include an accurate designation for northern China. This was a breakthrough in the tradition of T-O maps because it replaces the Garden of Eden with a concrete place in the tangible world where the Grand Khan of the Mongol empire resided. It is not surprising to see that this first breakthrough occurred in Armenia, because they possessed Prince Hayton's geographic account, which contained updated information about Asia. Moreover, the Armenian T-O map also contains contemporaneous information about Mongol-ruled Central Asia, such as Sarai.

These new pieces of information about the world would largely be incorporated into later medieval world maps, along with other new cartographic innovations. Some of these maps contain even more detailed information about Asia drawn from European travel accounts, such as those by Polo and Odoric. The most remarkable of these maps are the Catalan Atlas of the fourteenth century and Fra Mauro's works of the fifteenth century. These new types of world maps, for which we cannot find definitive precedents, added the continental contours adopted by newer cartography and included the Indian Ocean. The Catalan Atlas, produced in Catalonia, Spain, by a Jewish cartographer in 1375, is one of the earliest examples of this type of map. This unique map, which drew the world in four rectangular panels among other special features, is too complex to discuss in the space provided, so let us instead focus

---

17   The realistic contours of the Afro-Eurasian continents depicted in some of the T-O maps of late medieval Europe are similar to the earlier Sanudo map and Islamic circular world maps. For an example, see T-O map, Reims, 1417, in Edson (2007), p. 124.

18   The Armenian T-O map contains many of the traditional features of T-O maps, yet divides the continents by the Mediterranean and the Indian Ocean. (Galichian 2008). This new information – difficult as it is to trace its actual source – was little doubt influenced by knowledge about the Afro-Eurasian continents that had also influenced the Sanudo map.

on the Fra Mauro map in connection with the earlier circular world maps discussed above.[19]

The circular world map made in 1457–1459 by the Venetian Camaldolse monk Fra Mauro shares many features with the Sanudo world map, most notably, the contours of the Afro-Eurasian landmass and the division of the three continents by the Mediterranean Sea and Indian Ocean (see map D in Figure 8.1). In fact, when compared with the Sanudo map, Fra Mauro map's outlines reveal a greater resemblance to the Islamic maps that circulated during earlier centuries. While the Sanudo map placed the east on the top of the chart like other T-O maps, Fra Mauro's map, like the Catalan Atlas, placed the south on top, a typical convention among Islamic world maps. Several scholars, including Edson, Larner, and Morse, argue that this map demonstrates the influence of the ancient Greek geographer Ptolemy.[20] Ptolemy's map, reconstructed by contemporaneous Renaissance scholars, became popular in the fifteenth century; however, it is not a circular map like the one Fra Mauro created, nor does it place the south on top. Moreover, the Ptolemy map draws the Indian Ocean as a closed sea, a depiction that was outdated at the time of the drawing, as medieval Muslim geographers had already depicted it correctly as an open sea. David Woodward, who provides a detailed overview of the medieval world maps (*mappae mundi*) in *The History of Cartography*, also argues that portolan charts and Ptolemy's worldview influenced the specific land and sea division seen in Fra Mauro's map (which Woodward categorized as the transitional type of *mappa mundi*). However, the unprecedented land-and-sea division of the world already had appeared in the late thirteenth century, more than a century before the reappearance of Ptolemy's world geography in Europe. (Woodward 1987: 296–299) Piero Falchetta, who analyzed the contents of Fra Mauro's map in his book *Fra Mauro's World Map*, suggests possible Islamic influence based on the historical context in which the map was made, when contact intensified between Europeans and Muslims. (Falchetta 2006: 34) As the above discussion of the Sanudo world map demonstrates, the influence of Islamic maps dates back to the fourteenth century or earlier, and continued through its peak in the fifteenth century.

---

19 Its unique features include new cartographic styles influenced both by the European portolan tradition, including many compass rhumb lines, and by the Islamic cartographic works, that is, putting the south at the top, and new rich information about Asia based on the accounts of Marco Polo and John Mandeville. For a detailed discussion of this map, see Edson (2007), pp. 74–89.

20 Edson (2007), p. 140; Larner (1999), p. 148; Morse (2007), p. 28.

The incorporation of text makes Fra Mauro's map more novel and richly informed than the Sanudo map. His vivid description of Asia sets his map apart from any earlier map. Here, Fra Mauro used several sources, including a fifteenth-century European traveler named Nicola Di Conti (1395–1469), in order to update his own world map. There are also some sources that Fra Mauro apparently used but did not explicitly cite. This possibly includes works of Islamic geographic knowledge, which he would have gained through geographic works as well as information passed along to him from sailors. The other source that he undoubtedly used yet did not cite is Marco Polo's account, from which he gleaned information about eastern Asia. Fra Mauro locates China at the eastern end of the world, and identifies its two principal rivers, the Yellow and Yangzi rivers. Additionally, he plots nearly every town mentioned in Marco Polo's book, complete with annotations. For example, northern China is called "Chataio (Cathay)," and its capital "Chambalech (Beijing);" southern China is identified as "Regno Magno (i.e. Mangi)." On the coast lies "Chansay (Quinsai; Hangzhou)," which Polo refers to as "the finest and most splendid city in the world."[21] Further south lies the "*magnifico porto de Zaiton* (modern-day Quanzhou)," the most important port city in southern China for China's maritime contact with Indian Ocean networks. Falchetta has no doubts about the influence of Marco Polo's account on Fra Mauro's description of Asia, and indeed even argues that Mauro did not cite his source because, although it was famous, scholars did not acknowledge it at that time.[22]

There is little doubt that Marco Polo's account introduced the most striking new information about the world including Asia to Europeans during the late medieval period. Polo claimed that he served Khubilai Khan of China as a governor and wrote about what he experienced in person; however, no contemporaneous sources about Polo exist, and this has caused debates about his actual travels.[23] Even in his own time, Polo's fellow Venetians did not immediately accept his fascinating account, and some called him a liar. Even the famous author Dante (c. 1265–1321) did not cite or use Polo's content to describe world geography in his *Commedia*. (Schildgen 2002: 12–13) Nevertheless, Polo's book

---

21   Marco Polo, *The Travels of Marco Polo*, trans. Ronald Latham (London: Penguin Books, 1958), p. 213. Compare Marco Polo, *The Description of the World*, trans. A.C. Moule and Paul Pelliot (London: G. Routledge, 1938), p. 151.

22   Falchetta (2006), pp. 61–69. Scholars who compared Polo and the texts in the Fra Mauro map easily found that most of the contents about China in the map are identical to Polo. See also Cattaneo (2011), pp. 185–225.

23   For the most recent comprehensive summary of the Polo debates, see Vogel (2012), pp. 1–88.

grew popular throughout Europe. This, and subsequent accounts of Asia by authors like Odoric and John Mandeville (who copied from Odoric's account), may have inspired European interest in Asia. Circulation of these accounts contributed greatly to the spread of knowledge about Asia. It led cartographers seeking factual knowledge of the world (like the makers of the Catalan Atlas and the Fra Mauro map) to present new knowledge about Asia in visual form. This invention of the map based on a combination of old and new knowledge contributed to progress in Europe's geographic understanding of the world, and provides a clear piece of evidence verifying the crucial change in Europe made possible by its connections to Asia.

5   Conclusion

The people of late medieval Europe witnessed the proliferation of many different ideas about world geography that coexisted and conflicted with each other. This happened because of the conflicts between old ideas about the world and new ones that arrived from outside European society. This article has argued that the connections between Europe and Asia in the late medieval period made cartographic innovations possible. Through a number of political channels that gradually linked Europe to the eastern world in the thirteenth through fifteenth centuries – such as the Crusades and the Mongol empire – European geographers received new information that they used to revise their geographic conception of the world. This led to the production of new types of world maps, including the Sanudo map and the Fra Mauro map, which ushered in a new, realistic understanding of world geography.

The Sanudo world map was one of the first to introduce a realistic shape of Eurasian continents and seas to medieval Europe, which marked a major departure from the dominant T-O map form. The search for the source of this new cartographic innovation leads invariably to Islamic cartography, which most likely grew available through contacts with Muslims created by the Crusades and commercial connections. At the same time, a rich body of unprecedented accounts of Asia gradually traveled to Europe through Europeans who traveled to Asia through Mongol connections, like Marco Polo and many others. In his treatise, Sanudo, a contemporary of Polo, cited Hayton of Little Armenia yet did not incorporate new information about Asia from Polo's account. John Larner argues that Sanudo probably considered Polo's material to be "too complete a break" with the worldview represented by earlier T-O maps. (Larner 1999: 134) Indeed, it took several decades during the late medieval period for scholars and geographers to incorporate new geographical information into their works, because they had to check its reliability. The realistic depiction of the contours of continents that first appeared in the Sanudo world map was a path-breaking innovation; this new contour of the world would gradually become popular in the fifteenth century.

Travelers brought new and more detailed information about Asia to Europe, which soon found its way into cartographic works that helped Europeans to view the world in a novel way and to discern the map's important details at one glance. This is clearly seen in the Fra Mauro map of the mid-fifteenth century. Maps like this clearly demonstrate that the new information about Asia brought by travelers like Marco Polo gradually became incorporated into European geographic and cartographic works. This helped to facilitate a more concrete understanding of the wider world than ever before. It also inspired political leaders, scholars, and merchants to seek greater contact beyond Europe, which in turn led to unprecedented global interconnection through European voyages of exploration.[24]

Bibliography

Akbari, Suzanne Conklin (2008). "Currents and Currency in Marco Polo's *Devisement dou monde* and *The Book of John Mandeville*," in Suzanne Conklin Akbari and Amilcare A. Iannucci (eds.), *Marco Polo and the Encounter of East and West*. Toronto: University of Toronto Press, pp. 110–130.

Aujac, Germaine, et al. (1987). "Greek Cartography in the Early Roman World," in J.B. Harley and David Woodward (eds.) *The History of Cartography: Volume One: Cartography in Prehistoric, Ancient and Medieval Europe and the Mediterranean*. Chicago: University of Chicago Press, pp. 161–176.

Baynton-Williams, Ashley, and Miles Baynton-Williams (2009). *New Worlds: Maps From The Age of Discovery*. London: Quercus.

Cattaneo, Angelo (2011). *Fra Mauro's Mappa Mundi and Fifteenth-Century Venice*. Turnhout, Belgium: Brepols.

---

24   Of course, maps were not necessarily made to depict geographical exactness; often, they were made to reflect religio-political worldviews and ideologies. In this way, some maps with Christian ideologies were produced in Europe after 1500. For example, see the Clover Leaf Map made in 1581 by Heinrich Bünting, a protestant pastor and theologian, which places Jerusalem at the center of the three continents of Asia, Europe, and Africa that are drawn in the shape of a three-leaf clover, in Baynton-Williams and Baynton-Williams (2009), p. 38.

Dilke, O.A.W. (1987). "The Culmination of Greek Cartography in Ptolemy," in J.B. Harley and David Woodward (eds.), *The History of Cartography: Volume One: Cartography in Prehistoric, Ancient and Medieval Europe and the Mediterranean.* Chicago: University of Chicago Press, pp. 177–200.

Edson, Evelyn (2007). *The World Map, 1300–1492: The Persistence of Tradition and Transformation.* Baltimore: The Johns Hopkins University Press.

Falchetta, Piero (2006). *Fra Mauro's World Map.* Turnhout, Belgium: Brepols.

Galichian, Rouben. "A Medieval Armenian T-O Map," *Imago Mundi* (2008) 60(1): 86–92.

Harvey, P.D.A. (2006). *The Hereford World Map: Medieval World Maps and Their Context.* London: The British Library.

——— (1991). *Medieval Maps.* Toronto: University of Toronto Press.

Hayton, Frère (ca. 1235–ca.1314). *La Fleur des histoires de la terre d'Orient*, ed. C. Kohler (*Recueil des Historiens des Croisades, Histories Arménien*, vol. 2). Barcelona: Centre d'Estudis Medievals de Catalunya, 1989.

Jackson, Peter (2005). *The Mongols and the West: 1221–1410.* Harlow, England: Pearson Longman.

——— "The State of Research: The Mongol Empire, 1986–1999," *Journal of Medieval History* (2000) 26(2): 189–210.

Larner, John (1999). *Marco Polo and the Discovery of the World.* New Haven: Yale University Press.

Morgan, David (2007 [1986]). *The Mongols.* Oxford: Blackwell.

Morse, Victoria (2007). "The Role of Maps in Later Medieval Society: Twelfth to Fourteenth Century," in David Woodward (ed.), *The History of Cartography: Volume Three, Book One, Cartography in the European Renaissance, Part 1.* Chicago: University of Chicago Press, pp. 25–52.

Park, Hyunhee (2012). *Mapping the Chinese and Islamic Worlds: Cross-Cultural Exchange in Pre-Modern Asia.* New York: Cambridge University Press.

Polo, Marco (1254–1324). *The Description of the World.* Translated and annotated by A.C. Moule and Paul Pelliot. London: Routledge, 1938.

——— *The Travels of Marco Polo.* Translated by Ronald Latham. London: Penguin Books, 1958.

Safier, Neil and Ilda Mendes dos Santos (2007). "Mapping Maritime Triumph and the Enchantment of Empire: Portuguese Literature of the Renaissance," in David Woodward (ed.), *The History of Cartography: Volume Three, Cartography in the European Renaissance, Part 1.* Chicago: University of Chicago Press, pp. 461–468.

Sanudo, Marino [circa 1260–1343]. *Marino Sanudo Torsello, The Book of the Secrets of the Faithful of the Cross [Liber Secretorum Fidelium Crucis].* Translated by Peter Lock. Farnham: Ashgate, 2011.

Schildgen, Brenda Deen (2002). *Dante and the Orient.* Urbana: University of Illinois Press.

Sezgin, Fuat (2005). *Mathematical Geography and Cartography in Islam and Their Continuation in the Occident*. Part 1 (Being an English Version of Volume X of *Geschichte des Arabischen Schrifttums*). Translated by Guy Moore and Geoff Sammon. Frankfurt am Main: Institute for the History of Arabic-Islamic Science at the Johann Wolfgang Goethe University.

Skelton, R.A. (1958). *Explorer's Maps: Charters in the Cartographic Record of Geographical Discovery*. New York: Praeger.

The Periplus Maris Erythraei: Text with Introduction, Translation, and Commentary. Translated by Lionel Casson. Princeton, NJ: Princeton University Press, 1989.

Tibbetts G.R. (1992a). "Later Cartographic Developments," in J.B. Harley and David Woodward (eds.), *The History of Cartography: Volume Two, Book One, Cartography in the Traditional Islamic and South Asian Societies*. Chicago: University of Chicago Press, pp. 137–155.

——— (1992b). "The Balkhī School of Geographers," in J.B. Harley and David Woodward (eds.) *The History of Cartography: Volume Two, Book One, Cartography in the Traditional Islamic and South Asian Societies*. Chicago: University of Chicago Press, pp. 108–136.

Vogel, Hans Ulrich (2012). *Marco Polo Was in China: New Evidence from Currencies, Salts and Revenues*. Leiden: Brill.

Woodward, David (2007). "Cartography and the Renaissance: Continuity and Change," in David Woodward (ed.), *The History of Cartography: Volume Three, Book One, Cartography in the European Renaissance, Part 1*, Chicago: University of Chicago Press, pp. 3–24.

——— (1987). "Medieval *Mappaemundi*," in J.B. Harley and David Woodward (eds.), *The History of Cartography: Volume One, Cartography in Prehistoric, Ancient and Medieval Europe and the Mediterranean*. Chicago: University of Chicago Press, 286–370.

CHAPTER 9

Jamu: The Indigenous Medical Arts of the Indonesian Archipelago

Hans Pols

1   Introduction

For centuries, the oceans of this world have provided the most important ave-
nue for travel, exploration, trade, migration, diplomacy, and conquest as well
as for pilgrimages, proselytizing, and the dissemination of culture, wisdom and
knowledge. Discussing the geography and trade patterns of the Indonesian
archipelago, Adrian Vickers has stated: "Because Indonesia is a set of islands
linked through trade, it has always thrived on exchange, not only of goods,
but of ideas, cultures and languages." (Vickers 2005: 60) In addition to developing extensive networks of travel, trade, exchange and migration between
the thousands of islands of the archipelago, Indonesians established trading
connections with other areas in Southeast Asia, China, India, the Arab world,
and even with northern Australia. Arab merchants brought Indonesian merchandise to Europe while seafarers from Makassar brought the *trepang* or sea
cucumber to China. (Macknight 1976) With the establishment of these trading networks and the ensuing exchange of goods and merchandise came the
migration of individuals to and from the archipelago, the transfer of plants,
animals, and diseases (whether intentional or not), the introduction of new
products, production methods, religious beliefs (Hinduism, Buddhism, Islam
and, much later, Christianity), as well as the further dissemination of wisdom,
knowledge, and science. In this essay, the transfer and exchange of medical
wisdom and medicinal preparations to and from the archipelago are central.
In Southeast Asia, culture, religion, and knowledge followed trade routes,
which marked the main avenues of cultural contact and exchange. Indonesian
indigenous medical practices, which centred on the consumption of herbal
preparations, spread through these trade routes, which infused it, in turn, with
new ingredients and ideas. Indigenous Indonesian medical practices benefited
from the intensity of trade and exchange within and around the Indonesian
archipelago.

Jamu (the general name for a variety of indigenous Indonesian herbal
healing practices) has always been a dynamic and hybrid form of healing,

incorporating elements from many different medical traditions and sources, and therefore constantly developing and changing (there are several related healing practices, such as *pijat* or Indonesian massage). Throughout history, insights and skills on the preparation and application of *jamu* were communicated through oral traditions handed down within families, in particular by women. *Jamu* has a long tradition: images detailing the preparation of herbal medicine can be found on the mural sculptures of the Borobudur, a large Buddhist temple built in the 9th century located northeast of Yogyakarta. The local, oral, and dynamic nature of *jamu* poses distinct challenges for historians and anthropologists of medicine and healing practices intent on investigating it. In this essay, I will approach *jamu* by highlighting earlier attempts to gain an understanding of its nature, which provide an approximation of the art of *jamu* as it has been practiced for several centuries. Unlike the development of traditional Chinese medicine (which is documented in a great number of written sources), indigenous Indonesian herbal medicine has hardly received any historical or anthropological attention. It is the aim of this essay to highlight some of the characteristics of *jamu*, which is, albeit in a transformed form, still part of everyday life in Indonesia.

In this essay I will focus on *jamu* within the framework of the larger questions related to the development of science and knowledge in Asia. If we approach this question from more traditional perspectives within the history and philosophy of science, which privilege experimental scientific research conducted in laboratories over everyday practice, publications over oral traditions, researchers over practitioners, and scientific theories over tools, instruments, and practices, *jamu* can only appear as an inferior or preliminary form of medical knowledge which potentially could inform truly scientific research – an assessment already made by European physicians in the nineteenth and early twentieth century. Contrary to these approaches, I draw upon several new developments on the history and philosophy of science, medicine, and trade in order to develop a different framework to analyze the nature and significance of *jamu*. First of all, I analyze healing practices as a primary site for the development of knowledge instead of focusing on sites of abstract contemplation such as laboratories, universities, and other institutions of learning. Tinkering in everyday healing practices might inspire, at some point and under special circumstances, the development of abstract and written forms of knowledge, yet these everyday practices necessarily come first. Second, I emphasize the material or therapeutic aspects of healing practices such as *jamu*, that is: herbs, spices, and preparations, over the medical philosophies and spiritual worldviews that informed them. Third, I follow the recent emphasis on travel, trade, exchange, and the role of mediators in the histories of both trade and science, which explores the many relationships between both.[1] By following the global trade routes established before and during the era of exploration and imperialism, I analyze paths of trade, exchange and mediation, both of medical wisdom and herbal preparations. By following trade networks, it becomes apparent that European, Indian, Chinese, and Indonesian medicine have been connected through networks of trade and exchange which promoted the circulation of herbs, spices, medicinal preparations, medical lore, and wisdom. Rather than contrasting Eastern and Western medicine (both of which have often been described as more or less homogeneous and monolithic in nature, and almost always as following a separate logic of development), I propose to study the networks through which these different medical traditions have been connected throughout history from the beginning, and how these medical traditions have infused each other as specific times in history.

2   *Jamu* or Indigenous Indonesian Herbal Medicine

Jamu is a general designation of a set of healing practices that originated on the islands of the Indonesian archipelago and mostly consists of herbal preparations which takes advantage of the enormous biodiversity of the region. It was, and is, first and foremost practiced by women and overlaps in many ways with the preparation of food: many of the herbs and spices used in *jamu* are also used as seasoning (the close association between *jamu* and diet is not accidental; *jamu* practitioners see a close relationship between lifestyle, diet, health, and disease – as did most Western physicians before the 20th century). *Jamu* practitioners used to transmit their medical insights to their daughters, other family members, and friends orally; when literacy levels rose, collections of prescriptions were also composed and distributed. Based on a number of historical accounts and anthropological investigations of today's herbal healers, it is possible to form an impression of the art of *jamu* as it has been practiced over the last several centuries.[2] Central in *jamu* is the ginger family (*Zingiberaceae*): common ginger (Indonesian: *jahe*), turmeric (*kunir*), greater galangel (*lengkuas*), resurrection lily (*kencur*), *temu lawak, lempuyang pahit* ("bitter ginger"), and *lempuyang wangi* (the last three do not have common English names). The spices for which the Indies were known at the time of

---

1   In this approach, I build on recent developments in the history of colonial science. See, for example, Raj (2008), Roberts (2009), Schaffer et al. (2010) and Bala (2006).

2   For a popular overview of *jamu* today, with some comments about its origins, see: Beers (2001). See also Hans Pols (2010).

the Dutch East Indies Company (VOC) such as pepper, nutmeg, and cloves are also frequently used, as is a wide variety of fruit: papaya, pineapple, bananas, mangosteen, and others. At times, the bark of trees (for example, cinnamon) is used. *Jamu* generally consists of herbal preparations which have been boiled in water and strained. Traditionally, *jamu* is prepared fresh early in the morning, after which sellers, with baskets filled with bottles containing preparations on their back (*jamu gendong*) sell their wares on the street.

The ingredients used in *jamu* preparations illustrate its dynamic character: papaya and pineapple originated in Central America and were introduced by the Portuguese in the 16th century; coriander (or cilantro) came from southern Europe; *daun wortel* (the leafy part of carrots) was introduced by the Dutch. Galangal was prescribed by physicians in ancient India, Greece, and Arab countries (although it is not clear where it originally came from). Resurrection lily was used in Medieval European herbal medicine, after which it fell out of use and was only known in Asia. During the last few years, both echinacea and lavender have become popular ingredients in *jamu*, as have aspirin, steroids, and antibiotics, which are added by a few contemporary *jamu* sellers because of their reputation for restoring human health. (Lyon 2003) Even though the inclusion of these last few ingredients has been criticized by those who were looking for "pure" and "natural" medications, their use illustrates the inherent willingness of individuals who prepare *jamu* to use new elements that had been unknown to them before and which have been recommended by other healing traditions.

Indonesia has one of the most diverse eco-systems in the world. Because of the many islands, there are great variations in vegetation in the region. To both sides of the Wallace line (which was identified by Alfred Russel Wallace; it runs between Bali and Lombok, and between Kalimantan and Sulawesi), radically different flora and fauna can be found. This diversity has always provided ample resources for trade; not surprisingly, trade and exchange characterized the Indonesian archipelago long before Europeans arrived. Anthony Reid has stated that the many waterways made Southeast Asia unusually open to "seaborne traders, adventurers, and propagandists" because its sea-lines were not only ubiquitous, but also "remarkably kind to seamen." He called the region "uniquely favorable to maritime activity." (Reid 1993: 6) According to him, trade and commerce over the many waterways has been central to the way of life in Southeast Asia:

> Commerce has always been vital to Southeast Asia. Uniquely accessible to seaborne traffic and commanding the maritime routes between China (the largest international market through most recorded history) and the

population centers of India, the Middle East, and Europe, the lands below
the winds naturally responded to every quickening of international maritime trade. Their products of clove, nutmeg, sandalwood, sappanwood,
camphor, and lacquer found their way to world markets even in Roman
and Han times. (Reid 1998: 1)

Malay, a language which originated in eastern Sumatra and the area around
Malacca, became the language of trade and the markets of the region
(it became the primary language of Indo-Europeans and, later, the language of
Indonesian nationalists and the independent Republic of Indonesia). Already
during the late Middle Ages, Europeans benefited from the products of the
archipelago. At that time, condiments from the East became very popular with
wealthy families; the amount of spices added to one's food indicated one's
social standing. (Schivelbusch (1992) The immensely profitable spice trade,
long conducted through the Arab world, was taken over by the Portuguese, the
Spanish, and later by the Dutch and the English in an effort to increase profits
by decreasing the number of intermediary traders.

Patterns of trade established patterns for human migration and the
exchange of ideas. There has always been a strong influence from India on
the archipelago; undoubtedly, people from the Indian sub-continent brought
their medical traditions, among them ayurvedic medicine, with them. Chinese
individuals settled in the Indies before the arrival of Europeans; they brought
insights from Chinese traditional medicine and Chinese healers (*sinsehs*) with
them. During the colonial era, the Chinese came to dominate the local trade
in herbs, spices and indigenous herbal preparations; most pharmacies were
run by Chinese merchants and stocked herbs for both Chinese and Javanese
herbal medicine. Arab traders had brought Islam to the archipelago as well as
medical insights from the Greek and Arab worlds. Because the archipelago was
situated at the cross-roads of a number of trade routes, it was uniquely able
to receive and incorporate medical insights from a variety of sources. Trade
and exchange favored the development of knowledge locally; the inhabitants
of the Indonesian archipelago were in an unusually favorable position to take
advantage of this.

The basic philosophy of *jamu* is the idea that health is a balance between
the individual and his or her environment; this balance which can only be
maintained through prudent and moderate living, of which one's diet is an
essential part. A person's internal balance is mainly viewed within the polarities of hot versus cold and moist versus dry. If the body is getting too hot,
cooling herbs, spices, and food are given (such as cucumber); if the body is
getting too cold, heating equivalents are provided (ginger and hot peppers,

for example). In many Southeast Asian countries, it was believed that illness arose from excessive heat, a dangerous loss of heat, or the entry of excessive dry or moist air. (Reid 1992: 52) Health was related to the condition of one's *semangat* or life-force within; *jamu* or medical ministrations aimed to assist this inherently restorative life-force in case of illness. (Reid 1992: 55) The ideas behind the administration of *jamu* resemble basic notions in Chinese medicine, in particular the role of *qi* (life force), and *ying/yang* (hot/cold), although the theoretical ideas of Chinese medicine are much more elaborate. Viewing health as a precarious internal balance which is influenced by the relationship of individual to environment corresponds to ancient Greek ideas, which were transferred to the archipelago through Arab traders. Ayurvedic medicine contains similar notions. Rather than tracing the lines of influence between these medical philosophies in an attempt to locate their ultimate origin, I propose that the resemblances between these medical systems indicate that practitioners in all of them participated in larger networks of trade and exchange which transferred ideas and practices between different regions of the world. Apart from incorporating many elements that have come to the archipelago through trade and exchange, *jamu* is also embedded in the larger spiritual world of Java (which, in itself, contains many elements from abroad). For many practitioners, herbal medicine is essentially related to religion and the spiritual world – with the transfer of religious ideas, medical insights and ideas about maintaining health were probably transferred as well.

Because *jamu* has always been a local, hybrid, and dynamic form of medicine, it is difficult to give an account of its essence (and posit this essence as the "wisdom of the ancients," unchanging in nature, and informing *jamu* practice then and now). To provide such an account would require a great amount of creativity and an extensive use of what Johannes Fabian has called the anthropological imagination, which aims to remove any traces of Western contact from the image of primitive society it portrays, projecting back an image of an unspoiled, original, unpolluted, and, essentially a-historical essence, before it was touched by contact with outsiders, be they from the East or the West. (Fabian 1983) In the case of *jamu*, such an account would deny the dynamic, ever-changing, and persistently local nature of *jamu*. I have already mentioned that, throughout most history, *jamu* prescriptions were passed on orally; oral traditions are always inherently local and have an unusual capacity to incorporate new elements as new or unusual situations arise. (Ong 1982) Naturally, studying oral traditions poses distinct problems for researchers, in particular when they aim to study past oral traditions. It is only possible to gain an understanding of *jamu*, both its development and its dynamics, by analysing its appropriation by others, including physicians at the Sultanates, European physicians, and Indo-European women.

3   Representations and Appropriations of *Jamu*

To gain insight into the nature of *jamu*, I will briefly analyze a number of appropriations and concomitant transformations made by interested individuals. Of these I will discuss physicians at Javanese Sultanates, Indo-European women living in the major urban centers of the Dutch East Indies, and European physicians. These groups engaged in an exchange of ideas and medicinal preparations with the original preparers of *jamu*, and thereby carried the knowledge it embodied further – and, in the case of the last two groups, in ways that are easier to comprehend to readers familiar with Western medical traditions. One could argue that these chains of exchange, appropriation, and representation are not essentially different from those that were operative in the formation and transformation of *jamu* itself (which has consisted of exchange, appropriation and incorporation of other elements).

The various Sultanates on Java and in the Indonesian archipelago have always employed individuals renowned for their medical skills (from the late 19th century on, they also employed Western-trained physicians); these at times left written evidence of their insights. Some of these physicians aimed to provide an inventory of common medical insights; other attempted to refine them by further exploration, refinement, and at times unnecessary ornamentation. The Sultanates in Solo and Yogyakarta have large collections of these manuscripts, written on *lontar* (dried palm) leaves in traditional Javanese script. Some have been translated (mostly into Dutch).[3] The prescriptions in these books were undoubtedly inspired by the very rich *jamu* traditions in the region surrounding these Sultanates, but it should be taken into account that most court physicians embellished prescriptions to improve their taste, to make them appear unusual, and to impress their employers with their extensive knowledge. In addition, they needed to differentiate themselves from everyday *jamu* peddlers. Court physicians were also able to read texts from Indian medical tradition and probably incorporated these medical insights in their writing. Because they were literate, they had more avenues to acquire

---

3   See, for example, van Hien (1924) and Smits (1928). De Visser Smits was better known as one
    of the most prominent freemasons in the Dutch East Indies. The freemasons were unusually
    interested in Eastern wisdom.

medical insights from elsewhere – their wisdom was based on a variety of sources, of which *jamu* as it was practiced on the streets nearby was only one. When carefully analyzed, these early written sources do provide insights into early Javanese medical ideas and practices.

In the 17th and 18th century, when European traders arrived in the Indies, the physicians who accompanied them eagerly collected herbs, spices, and local medical lore, which they appreciated highly and documented extensively. The writings of Garcia de Orta (1501–1568, Portuguese physician working in Goa),[4] Bontius (Jacob de Bondt, 1592–1831), physician in Batavia),[5] Hendrik Adriaan van Rheede tot Drakenstein (1636–1691, Dutch physician and botanist, worked at the Malabar coast, India),[6] and Rumphius (Georg Eberhard Rumph (1627–1702), German physician, worked on Ambon, nicknamed the Pliny of the East)[7] provide fascinating insights into indigenous herbal practices. All of these accounts were highly positive about the herbal medicine of the East. Bontius even stated:

> Every Malayan woman practices medicine and midwifery with facility; so (I confess that it is the case) I would prefer to submit myself to such hands than to a half-taught doctor or arrogant surgeon, whose shadow of education was acquired in schools, being inflated with presumption while having no real experience.[8]

Van Reede tot Drankenstein assembled his *Hortus Malabaricus* with the aid of Ezhava physician-botanists rather than Brahmin scholars, who relied too much on written sources to be of any help. Richard H. Grove argues that both de Orta and van Reede's writings were "profoundly indigenous texts" because they extensively relied on local informants. (Grove 1996: 126) Both texts deeply influenced European botany; in particular van Reede's *Hortus* influenced Linneaus, who extensively consulted it when he studied at Leyden. Many of van Reede's insights (which were, in many ways, insights of the Ezhava) can be found in Linneaus' writings later on.

---

4 Orta (1563). Orta's work was the first one to provide insights on medical ideas from India and provided one of the first overviews of medicinal plants from that area.

5 On Bontius see Cook (2005, 2007).

6 Heniger (1986); Grove (1995, 1996).

7 Beekman (1981), and Georgius Everhardus Rumphius, and E.M. Beekman (1999 [1705]), Rumphius (2011 [1741]), vol. 1–6.

8 Bontius, *De Medicina Indorum*, quoted in: Cook (2005), p. 100.

The writings on the medicine of the Indies (broadly defined, as it was back then, including India, Sri Lanka, and Indonesia) were widely read in Renaissance Europe. Insights from the East as reported in these works, combined with the spice trade, which made herbs and spices from the East available in Europe, and the establishment of botanical gardens near universities, where attempts were made to grow medicinal plants from the East, inspired a medical Renaissance in Europe. In particular the herbal garden established by Carolus Clusius at the University of Leyden in 1594 and the clinical teaching of Hermannus Boerhaave propelled Leyden as one of the leading centers of medical teaching in Europe. (Cook 2007) In his path-breaking work on botany, medicine, and colonialism in the Low Countries, Harold Cook relates the Dutch medical Renaissance to the spice trade of the VOC and the impulse it gave to careful investigation of botany, which aided in the recognition of valuable merchandise and potentially useful medicinal plants. According to him, the detailed description of natural objects was closely related to the development of colonial trade networks. During the Renaissance, European botany and medicine (which were, at that time, hardly separated) incorporated many insights of Eastern medical practitioners and botanists – but often without attribution. These insights had come to Europe through trade and the writings of explorers and physician/botanists. Even though Western medical knowledge tended to define its own identity in contrast with Eastern superstition, religion, and ossified tradition, the East actually contributed many elements which were obtained by Western-trained physicians through the many relationships of trade, exchange, and creative appropriation which were essential to the development of medical knowledge in the early modern era.

A third group of sources derives from a small group of generally city-based Indo-European women who studied indigenous herbal medicine for their own purposes. The urban centres of Java became strongholds of Indo-European culture when male soldiers and merchants, after completing their obligations to the VOC or the Dutch colonial government, decided to remain in the Indies and formed relationships with indigenous women.[9] In many ways, Indo-Europeans became intermediaries between Europeans and the indigenous population. Indo-European women benefited from traditions of herbal medicine that were part of their Indonesian heritage or consulted indigenous women. From the early nineteenth century, some of these women collected written recipes and compiled them in books; a few of these survive because they were published.[10]

---

9   On Indo-European communities in the Dutch East Indies see Taylor (1983); Bosma, and Raben (2008).

10   van Gent-Detelle (1880 [1875]), van Blokland (1899).

Most prescriptions in these books merely list ingredients, with some indication of the amounts that were used, and provide rudimentary information on how to prepare them. In other words, the authors of these books assumed that their readers had most of the necessary insights for preparing *jamu* so that it was not necessary to spell out all the details. The insights and skills these Indo-European women possessed were already somewhat removed from the Javanese women who prepared *jamu*. In their writings, broader cosmological, spiritual, and religious considerations are completely absent – all attention is focused on ingredients and the conditions, partly defined in European medical terms and partly following traditional insights, which they aimed to cure.

These Indo-European practitioners of herbal medicine forged a hybrid healing tradition incorporating elements from *jamu* and Western medicine. They generally incorporated ideas about the nature of disease and diagnosis from Western medicine, while appropriating *jamu* formulations within the domain of treatment. They attempted, for example, to cure tuberculosis with a specific collection of herbs, colds with ginger, fevers with cucumber, etc. The Western approach to the nature of disease reified them as individual entities (supposedly located in specific organs) rather than viewing them as the outcome of an imbalanced state of the organism. In their approach, *jamu* was transformed in a way that made it resemble Western medicine by following the assumption that there was a one-to-one correspondence between disease and medication. Not surprisingly, the ideas of these women were far more accessible to European physicians who were interested in indigenous herbal medicine in the 19th century; the latter obtained most of their information from these women rather than from *dukuns* (traditional healers) and other indigenous *jamu* producers. (See Pols 2009) Some of these Indo-European women became formidable competitors for European physicians, as most city-dwellers would consult the former rather than the latter. (van der Burg 1884)

European physicians working in the Dutch East Indies during the 19th century constituted a fourth group eager to explore Indonesian herbal medicine – often in order to enhance their pharmacopeia. After all, they were painfully aware that they were not able to treat even the most common of diseases in the Indies which somehow did not pose a problem for traditional medical practitioners.[11] The investigations of European physicians provide us with detailed information on the ingredients used in *jamu* and their modes of preparation. Interestingly, the way these physicians acquired their information was closely related to trade and exchange: they visited *pasars* (markets) where herbs and spices were traded, as well as pharmacists (who generally

---

11   For an overview of the "medical market" in the Dutch East Indies see Hesselink (2010).

were Chinese but sold herbs used for the preparation of *jamu* and *jamu* preparations in addition to Chinese medications). They also gained information from Indo-European women, although most of them were not eager to share their medical insights out of fear that it would erode their market position. (Pols 2009: 178–180) A small number of Western-educated pharmacists were employed at the Buitenzorg [Bogor] botanical gardens to analyze *jamu* to isolate effective ingredients. W.G. Boorsma was the most important of these; he built up an extensive network of informants to guide his research. (Boorsma (1913)

Knowledge about indigenous herbal medicine was appropriated from illiterate and semi-literate Javanese women who prepared and sold *jamu* to physicians at the Sultanates as well as to European physicians, botanist, and pharmacists. In the process, *jamu* became alienated from its original source and context, in which spirituality and a wide range of animistic beliefs were central. At the same time, the level of abstraction increased. Traditionally, this movement has been viewed as an improvement as far as the scientific level of this knowledge is concerned. One could also view it as a process of alienation from the original site of creativity, experimentation, and knowledge production. When knowledge enters books it has the potential to become rigid. Van Reede tot Drakenstein consulted a group of physicians-botanists rather than Brahmin scholars for his *Hortus Malabaricus* because he was convinced that they were better informants (the Brahmins just quoted their books, he stated). At times, increased abstraction and distance from the original practitioners decrease the amount of insight one can gain. Nevertheless, it is essential to analyze the flow of information from its source to later recipients, and the transformation this knowledge underwent in the process – in particular since the origins of this knowledge is often erased from later accounts. This account would become even more dynamic when the original informants are not seen as ultimate sources but only as nodes in larger networks that have connected medical traditions for long periods of time.

At this point, I would like to introduce the concept of the *desa* physician (*desa* means rural village), which is an adaptation from the concept of peasant intellectual introduced by Steven Feierman. Feierman uses this concept as an antidote to the asymmetry in which we tend to view the ideas articulated by the educated European elite and those by illiterate indigenous individuals. He analyzed the way peasants in rural Tanzania discussed social change "with eloquence and penetration." (Feierman 1990: 23) In Feierman's definition, intellectuals are defined though their participation in "socially recognized organizational, directive, educative, and expressive activities." (Feierman 1990: 17–18) Feierman focused on the way peasants discussed social and political changes related to decolonization, land-ownership, and leadership. His definition can easily be modified for the purposes of this essay: a *desa* physician is someone who, when faced with a specific condition of ill-health, is able to select ingredients, prepare a medicinal concoction, and apply it as a means to alleviate that condition and to facilitate the restoration of health. *Desa* physicians are able to do this by astutely drawing a range of ideas, insights, and healing practices common to a generally shared healing culture or to smaller groups or families. In addition, *desa* physicians are able to incorporate new elements, be they herbs, spices, plants, or anything else, as well as new ideas and approaches, in their healing practices. A basic and flexible philosophy (based on the polarities hot/cold and moist/dry) structures the thinking of *desa* physicians. Yet, in Western appropriations of *jamu*, the ingredients that were used were central while the basic philosophy is mostly lost. In proposing the concept of the *desa* physician, I attempt to provide an alternative to dichotomies which are still used in the history and philosophy of science, including knowledge/tradition; science/application; science/folklore; scientist/(lay) practitioner (and, overlapping many of the previous distinctions, male/female). By viewing the individuals who develop, prepare and apply *jamu* as *desa* physicians, I emphasize their ability to engage in healing practices that are informed by a more or less explicit set of ideas as well as their ability to modify that practice by incorporating new elements and ideas. Their practices tend to change and improve as a consequence of increased trade and exchange with others – not only other *desa* physicians, but also traders, travelers, and practitioners from other medical traditions. In other words, as pragmatic and curious practitioners, they are not unlike physicians and healers in many other parts of the world and at different moments in history.

4   *Jamu* as Knowledge

In this essay, I examine *jamu* to explore conceptions of what constitutes science and question earlier distinctions between science and "other forms of knowledge." I propose to explore medical practices when contemplating the development of knowledge and science and, in addition, to emphasize the material elements of science over theories, ideas, and rationalizations. Following recent approaches in science studies, in particular Bruno Latour's thoughts on what enables scientific knowledge to "travel" to locations other than the one it originated in, and Peter Galison's views on trading zones, I will propose an approach which emphasizes that medicine (or other knowledge traditions) would thrive when it incorporates material elements from other locations as an outcome of trade and exchange. Because of the existence of extensive trade networks, it can be expected that, at the time of the European dark ages, many medical traditions incorporated elements from other traditions. In many ways, the medical traditions of the European and Arab worlds, India, China, and Indonesia were part of larger networks through which elements were transferred and exchanged on an almost continuous basis.

4.1   Medicine as a Model of Science

Science has generally been defined as an esoteric body of knowledge, ideally presented in quantitative statements, produced by individuals employing unique and special methodologies. The distinguishing character of scientific knowledge is that, although it is conceived, developed, and verified at unique locations, its validity transcends time and place. Initially, philosophers have focused on scientific theories, which are generally presented in written or mathematical form. Traditionally, historians of science (and, in addition, philosophers of science) have preferred to focus on the history of physics – the queen of the sciences, a science without the complications associated with an affiliated practice, as it was initially focused on the prediction of the movements of the planets. This traditional image of the nature of science is complicated if we consider medicine, because it has always developed in combination with caring practices in addition to its engagement in investigative pursuits. To designate the latter as a site for the development of medical knowledge and the former as a one for mere application does not do justice to the many ways in which both are interrelated. The practice of medicine has existed for a long time, while medical research conducted independently of medical care has only recently developed. Practice (in medicine, or in whatever craft) is always primary; under specific conditions a site can develop from these practices to focus exclusively on investigative pursuits. Yet practice always comes first.

If we focus on the development of scientific knowledge, characterized by the intermingling of ideas, insights and practices, then it could easily be argued that medicine was the very first of all sciences, as human beings have always had an interest in ways to alleviate the symptoms of disease, restore health, and increase strength and vigor. Through trial and error, random experimentation, observation, and communication, human beings explored the natural world for means, first, to feed themselves and to stay alive and, second, to alleviate symptoms of disease. Within several folk medical practices arose moments of reflection – on the nature of the human body, the characteristics of disease, and the best ways to alleviate symptoms – which constituted the beginnings of medicine as an investigative and reflective practice. The distinction between research and practice as well as the creation of groups of individuals who devote themselves exclusively to research have appeared only recently within the history of science and medicine. To focus on physics as a science without clear applications might be philosophically satisfying but overlooks much larger and widespread sets of practices undertaken by groups of individuals who systematically reflected on health, disease, and the nature of the human body, reflections which were immediately relevant for the societies in which they lived. Physicians have always, and everywhere, outnumbered physicists.

Naturally, as a historian of science and medicine, one could wonder why some medical practices developed more extensive reflective practices which became the domain of specialized or elite healers. These healers naturally appropriated existing healing traditions and enhanced their insights by acquainting themselves with a variety of healing traditions. Through verbal or written exchanges between such elite physicians, medical knowledge potentially increased. One could hypothesize that the strength of medical traditions is related to the immediate natural resources at one's disposal, the availability of other resources obtained through trade and exchange, the ingenuity (and audacity) of individuals willing to engage in medical experimentation, and the number of medical insights available from other healing traditions. In particular, the exchange of ideas beyond one's own cultural circle can be expected to benefit medical insights. Seen in this way, Western medicine does not differ significantly from medicine as it is practiced elsewhere in the world.

4.2 Trade and the Dissemination of the Material Elements of Medicine

I have referred several times to the importance of trade, trading routes, and networks of exchange. I would now like to make a small detour to the work of Peter Galison on microphysics, in particular his introduction of the concept of the trading zone, which he uses to analyze how communication and transfer of knowledge is possible in a highly fragmented and disunified field of science – a field of science in which different participants find it hard to communicate with each other but are nevertheless able to collaborate on specific research projects. I would like to take his ideas literally and place them back within the specific context from which they were taken (an anthropological account of trade relations in a non-Western context). In other words, I would like to apply this idea to the times when trade and exchange went hand in hand with the formation of unique medical traditions in Southeast Asia, Europe, and the Arab World. According to Galison, the exchange of tools and experimental technologies, rather than ideas, can take place in trading zones, which are inherently local and need to be negotiated by participants from both sides. Often, as in the marketplace, the items exchanged have radically different meanings for donor and recipient (buyer and seller), but this does not impede trading (in many respects, it can be said that stripping objects and tools from excess meaning enhances trading). From Galison's perspective, the art of trading can be defined as the ability of two scientists (or two groups of scientists) to meet temporarily in a specific location to which they bring rather different expectations, goods, and tools while being able to negotiate an exchange that both parties consider sufficiently valuable. To quote Galison: "What is crucial is that in the highly local content of the trading zone, *despite* the differences in classification, significance, and standards of demonstration, the two groups can collaborate." (Galison 1999: 146) Only a consensus to trade and exchange is necessary despite the presence of "two vastly different symbolic and cultural systems, two perfectly incompatible valuations and understandings of the objects exchanged." (ibid.) Nevertheless, it is necessary for both parties to develop trust in each other to continue to engage in trade and exchange, an element Galison does not explore further.

Communication and mutual understanding are not of central importance in Galison's trading zones – scientific theory therefore has very little to do with it, although it should be emphasised that, as a result of activity in the trading zone, experimental work on both sides can continue in new and creative ways, which, ideally, will result in theoretical and experimental innovation. This expected outcome motivates trading but does not, in itself, contribute to it. Trading benefits from a relative lack of concern about communicating on a *theoretical* level; it focuses almost exclusively on the exchange of *material* elements. In a chain of trades (which is very common in the context from which Galison lifted the idea of the trading zone), the initial donor and the ultimate recipient do not even meet as they are geographically separated; their ideas about the nature of the exchange have become irrelevant. To enable trade to proceed, only a minimal form of communication is necessary. Using an example from Papua New Guinea, Galison asserts that a contact language, a local pidgin, is "construed with the elements of at least two active languages" to facilitate trading. (Galison 1999: 153–54) This newly created language, which often incorporates simplified concepts and constructs from both sides, enables communication to facilitate trading. At some point, a group of individuals will inhabit the trading zone; for these individuals, this newly created language becomes their first and natural language. These inhabitants or mediators can be called the creoles of the trading zone.

In his account, Galison emphasizes relatively brief encounters, yet in the world of trade and exchange, long-term relationships in which trading partners can develop trust are essential. Trust arises in long-term exchange relationships, in particular the desire to continue this relationship. In the exchange of medical insights, other contact zones which contain long-term relationships of trust were significant as well. In the Dutch East Indies, many European physicians acquired insight into the nature of *jamu* after marrying a local, Indo-European, or indigenous woman. (Pols 2009: 180–186) Others asked their patients about the treatments they had received previously or questioned their own servants (Garcia de Orta's female slave was an important source of information; she often appears in his *Colóquios*).

Galison's analysis highlights several points that can be applied to the analysis of *jamu* as a form of knowledge – in particular by taking the emphasis on trade and exchange seriously. First of all, Galison encourages us to take an active interest in the *material* elements of science, which, after all, make trade and exchange between different groups possible. Bruno Latour has demonstrated that it can be fruitful to analyze the material culture of science (tools, machines, instruments) rather than ideas, verbalizations, and theories. (Latour 1987) Secondly, Galison emphasizes the importance of pidgin – a hybrid language which is limited in nature but which, nevertheless, makes exchange possible. We could extend his analysis by asking, in a more general way: which factors facilitate trading and which factors impede it. These factors determine the transfer of the material elements of science and medicine or the embodied elements of knowledge. In the case of *jamu*, medicine in Southeast Asia, and medicine in general, material elements or goods (like herbs, spices, but also medical preparations) can be transported over much longer distances than ideas, theories, and explanations. Because of the vast differences in medical traditions and the different languages in which they are expressed, extraordinary effort would be required to translate medical theories and philosophies from different cultural zones. Some medical insights are transferred much more easily, however. Naturally, traders are interested in the nature of the goods they transport and are eager to gain insights into their various uses (culinary, medical, and otherwise). Nevertheless, ideas and theories that explain the use of one's medicinal merchandise are not transferred as easily and as far as the goods which they are originally related to, in particular when exchanges involve a chain of buyers and sellers that transfer goods to distant markets in vastly different cultural spheres.

Goods are transferred along these trading chains relatively unchanged (traders are of course interested in ways to make their goods transferrable by avoiding spoilage and decay – goods that remain in good condition for longer periods of time are more suitable for trading). Naturally, only the simpler, easier, and more straightforward types of ideas get transferred as well – if the ideas presented by the original donor were highly intellectual and well-reasoned, one can expect that, after a chain of transfers, only the basic outline of these ideas remained. So in this model on the circulation and transfer of medical knowledge, less complex and more straightforward types of understanding travel farther and are less likely to be modified in the process. The systematic corrosion that exists in chains of transfer leads towards simplification rather than to elaboration and confabulation (except, may be, at the hands of eager traders who wish to inflate the value of their goods symbolically). After all, the transfer of ideas and insights has to occur in a *pidgin* that ties donor and recipient (or buyer and seller) together in the highly local context of the market. When Latour discusses immutable mobiles, or inscriptions that remain unchanged despite their transfer to a different context, he was approaching the same issue – long distance communication – from a different angle. (ibid.) He wondered how linguistic expressions (or diagrams, maps, charts, numbers) could retain their meaning by becoming less dependent on specific contexts. This, naturally, is a central question in science studies, and is relevant in the dissemination of scientific knowledge. I wish to complement Latour's approach by analysing the transfer of goods that can travel almost independently of their meaning – or, in other words, acquire different meanings in different contexts. These goods have the potential to spread medical insights or, at least, the *material* elements of medicine, but only because no attempts have been made to exert long-distance control or to establish a center of calculation. In the case of herbs and spices, merchants asked themselves how they could increase the stability of their goods – to prevent them from spoiling, drying out, becoming moldy, because the chain of trade could only be as long as the time over which the original nature of their goods would be preserved. But they did not aim to control the *meaning* and *uses* of their goods. These are, in the end, dictated by the framework of the recipient.

In the transfer of knowledge, it is essential to focus on the transfer of the material culture in which this knowledge is embodied: the goods, products, and merchandise. The flow of goods is enhanced when the value attached to these goods increases, and when these goods can be stabilized so as to not lose the highly valued characteristics present at the time of initial purchase. Of second importance is to focus on the relatively simple and straightforward ideas that accompany the elements of this material culture of knowledge, or, at least, the ideas and insights that are present at the final end of the chain of transfer and trade. Of least importance are the highly elaborate, well thought-out ideas of court physicians, sages, priests, shamans, and those individuals who are recognized within the society where the goods originate as highly learned. However impressive the ideas of these individuals are, they are much harder to transfer successfully over long distances and to different cultural contexts. In other words, to look at the circulation of knowledge practices in the ancient and early modern world, focus has to be on *materia medica* and relatively simple medical ideas, not on philosophers, sages, court physicians, or the medical researchers at the universities of Renaissance Europe.

The material elements of medicine – herbs, spices, food items, and medical preparations – were part and parcel of the trade within the Indonesian archipelago and within Southeast Asia. Individuals possessing the art of preparing and administering *jamu* benefited from these trading networks – both with respect to the material elements used in their preparations and broader ideas about health and disease that traders brought with them. In turn, it can be expected that *jamu*, in many of its forms and varieties, contributed to medical approaches within the region and beyond. Through the extensive trading networks of Chinese and Arab traders, elements of *jamu* were transferred to Europe, inspiring a medical Renaissance there. It is therefore more useful to analyze *jamu* as a hybrid form of medicine, incorporating elements that became available through trade and exchange, instead of searching for its deeper essence that can be posited to precede exchange and the incorporation of foreign elements. In many ways, it is probably more fruitful to analyze medical traditions in the region (China, Japan, and India) and distant regions (the Arab world and Europe) along the lines suggested in this essay by investigating how and to which extent they incorporated medical folk remedies as well as goods and ideas acquired through trade and exchange. Trade between East and West has existed for a very long time. One can therefore expect that elements of the medical traditions in Europe and Asia have been traded and exchanged for a long time as well. It would therefore be interesting to investigate the many linkages between these medical traditions and to analyze them as part of a world-wide network of medical practices linked through trade and exchange. Emphasis would then be given, first of all, to the circulation of the material elements of medicine and, secondly, to the exchange of basic medical ideas.

The effect of trade and exchange between East and West on the development of Western science should not be underestimated. Harold Cook, in *Matters of Exchange*, points out that the rise of modern science took place in the first age of global commerce and that an economy focused on trade and exchange "changed the terms of reference for intellectual investigation." (Cook 2007: 1–2) According to Cook, the interest in trade and the goods acquired in the East inspired a curiosity about matters of fact and the exact description of objects, which had, until that time, not been part of European intellectual traditions. Alix Cooper has argued that the many new species described on voyages of exploration inspired Europeans to develop a systematic approach to botany. Europeans started to observe nature around them for the first time after seeing depictions of unusual plants from other parts of the world. (Cooper 2007) In addition to this observational approach to the natural world, the voyages of trade and exploration also brought goods that were used to good effect in European medical practices, as well as a number of pioneering guides about the medicinal plants and spices of the Indies. In many ways, the goods from the East in addition to Eastern insights that had been appropriated by European observers inspired Europe's medical Renaissance.

5   The Appropriation, Transformation, and Transfer of *Jamu*

In the past (and to some extent even today), traditional Javanese healing practices were conducted within the framework of Javanese cosmology, which is suffused with supernatural elements, the presence and often unpredictable influence of spirits, and a variety of magical powers that can be harnessed by following the directions of traditional healers (*dukun*).[12] Their prescriptions contained elaborate instructions on the time and date when specific infusions should be prepared and administered; these corresponded to auspicious dates in the Javanese calendar. Preparation required the uttering of specific incantations or prayers. A common theme in Dutch colonial novels was the demise of the European protagonist at the hands of the Javanese concubine he had abandoned. She then ensorcelled him by placing under his pillow pulverised beetles that had feasted on leaves of the *ketjoeboeng* flower (*Datura fastuosa*, a plant with known hallucinogenic properties).[13] After this "treatment", he would fall under her spell and return to his former concubine. The actual ingredients of the potions or ingredients provided by a *dukun* were only part of a larger set of prescriptions which derived their meaning from the place they had within the Javanese cosmological order; a *dukun*'s interventions aimed to restore imbalances and disturbances in that order.

In everyday Javanese life, physical and mental ailments were, first of all, interpreted in a common sense perspective in which health was determined by the balance between hot and cold, and moist and dry. *Jamu* prescriptions aimed to restore this imbalance by providing a prescription that would counteract reported symptoms. It is doubtful whether the women who sold *jamu* on the street were fully cognisant of the elaborate cosmology of traditional healers, even though these healers shared the view that health is a balance of a variety of elements and forces. *Jamu* sellers prepared standard infusions which

---

12   van Hien (1986). See also Wiener (1995), and Wiener (2003).
13   Wiener (2007), p. 506. Wiener partly bases herself on van Hien.

aimed to help individuals to maintain their health rather than addressing specific ailments. Women who prepared *jamu* at home used the same ingredients and applied them on the basis of the same common sense approach to health.

The few printed sources containing *jamu* prescriptions that appeared during the late nineteenth century contained lists of ingredients for diseases which were described in both Western diagnostic terms such as fever, pneumonia, or the flu as well as the common sense Javanese framework which defined disease as imbalance.[14] These books were written in Malay by Indo-European women who aimed to make Javanese medical insights available to literate readers and contained no references to supernatural forces, spirits, and spells. In works published during the first part of the twentieth century, disease conditions were increasingly described in Western medical terminology only; the infusions that alleviated them were given as lists of ingredients and included directions on how to prepare them. The most influential book in this genre was *Guidance and Advice regarding the Use of Indies Plants, Fruits, etc*, written by Mrs. J. Kloppenburg-Versteeg, which was written in Dutch.[15] It listed the names of plants and herbs used in her prescriptions in Javanese, Malay, and Latin. To help readers of her manual to prepare her prescriptions, an accompanying volume contained illustrations of all plants which were used in her prescriptions. In this manual, a Western understanding of health and disease predominates, although the Javanese common sense perspective is still present. References to spirituality, cosmology, and the supernatural are not made as these were not relevant to this author and her readers. Recipes were given as cures for specific disease conditions, defined in Western medical terms.

European physicians who explored indigenous *materia medica* in the sixteenth and seventeenth century rarely met with traditional healers (Van Reede tot Drakensteijn was an exception). Language barriers and a lack of understanding of Javanese cosmological beliefs made these interactions exceedingly difficult. They generally gained their information from slaves, their household personnel, Chinese pharmacists, and, in some cases, their wives. One of the most common ways to acquire insights into the medicinal properties of plants was going to the local market; sellers of herbs and spices were eager to enlighten potential customers. When plants, herbs, and spices were shipped to Europe, only rudimentary elements of the elaborate views of traditional healers and the more basic ideas most urban dwellers held about their proper use were transmitted. Those physicians who could visit botanical gardens

---

14   Van Gent-Detelle (1880 [1875]). Van Blokland (1899).
15   Kloppenburg-Versteegh (1934 [1911]).

where plants from the Indies had been grown successfully often knew little about the ideas of the traditional healers who might have used these plants in their own prescriptions. They rarely tried to understand the views of these healers as they were not interested in understanding different cultural perspectives. While it was difficult to transport plants, herbs, and spices from the East Indies to Europe without loss of taste, spoiling, or other forms of decay, transferring ideas about health and healing between those two different parts of the world was even more difficult.

6   Conclusions

In this essay, I propose to reframe the analysis of the development of science in a world that has always been global by focusing on the practice of medicine rather than, for example, physics; by focusing on practice rather than representations; by focusing on the material elements of healing practices rather than the wisdom that informed them; and on women rather than men. But, most importantly, I have focused on the importance of contact zones, places where individuals of different backgrounds and nationalities meet and interact, often through creative misunderstanding. In this approach, I build on the work of Lissa Roberts, who has emphasized the importance of highlighting "the productive role played by globally situated intercultural exchanges in the history of science and history more generally, while simultaneously recognizing the asymmetrical character of the conditions that often attended such encounters" (Roberts 2009:10) as well as on the work of Kapil Raj who has emphasized that encounters are "complex historical events and moments of discovery" and that the "worlds of trade and learning were closely intertwined." (Raj 2008: p. 8 & p. 18) The focus on medicine as a practice enables us to focus on one of the more important precursors of later intellectual and experimental practices that are central to modern science. The emphasis on the material aspects of medical cultures, in association with an emphasis on trade and exchange, has the potential to provide insights into factors that have stimulated such practices throughout history. The perspective developed in this essay explains the development of indigenous herbal medicine in the Indonesian archipelago by referring to the abundant natural resources of the region and the extensive trade networks that spanned it. It also encourages us to explore the many ways in which medical traditions in different regions of Europe, the Middle East, and Asia have been connected through networks of trade and exchange. In particular, explorations in the history of pharmacy can be expected to shed light on the nature of these networks. After all, herbs, spices, and medicinal preparations can be expected to travel further than medical ideas. It is by following the material culture of medicine that linkages between the medical practices of different cultures could be established.

Bibliography

Bala, Arun (2006). *The Dialogue of Civilizations in the Birth of Modern Science*. New York: Palgrave MacMillan.

Beekman, E.M. (1981). *The Poison Tree: Selected Writings of Rumphius on the Natural History of the Indies*. Amherst, MA: University of Massachusetts Press.

Beers, Susan-Jane (2001). *Jamu: The Ancient Indonesian Art of Herbal Healing*. Singapore: Periplus.

Boorsma, W.G. (1913). *Aanteekeningen over Oostersche Geneesmiddelleer Op Java* [*Notes on Eastern Herbal Medicinal Arts on Java*]. Buitenzorg: 's Lands Plantentuin.

Bosma, Ulbe, and Remco Raben (2008). *Being "Dutch" in the Indies: A History of Creolisation and Empire, 1500–1920*. Translated by Wendie Shaffer. Singapore: National University of Singapore Press.

Cook, Harold J. (2005). "Global Economies and Local Knowledge in the East Indies: Jacobus Bontius Learns the Facts of Nature," in Claudia Swan and Londa Schiebinger (eds.), *Colonial Botany: Science, Commerce, and Politics in the Early Modern World*. Philadelphia, PA: University of Pennsylvania Press, pp. 100–118.

——— (2007). *Matters of Exchange: Commerce, Medicine, and Science in the Dutch Golden Age*. New Haven, CT: Yale University Press.

Cooper, Alix (2007). *Inventing the Indigenous: Local Knowlege and Natural History in Early Modern Europe*. New York: Cambridge University Press.

Fabian, Johannes (1983). *Time and the Other: How Anthropology Makes Its Object*. New York: Columbia University Press.

Feierman, Steven (1990). *Peasant Intellectuals: Anthropology and History in Tanzania*. Madison, WI: University of Wisconsin Press.

Galison, Peter (1999). "Trading Zones: Coordinating Action and Belief," in Mario Biagioli (ed.), *The Science Studies Reader*. London: Routledge, pp. 781–844.

Grove, Richard H. "Indigenous Knowledge and the Significance of South-West India for Portuguese and Dutch Constructions of Tropical Nature," *Modern Asian Studies* (1996) 30(1): 121–143.

——— (1995). *Green Imperialism: Colonial Expansion, Tropical Island Edens and the Origins of Environmentalism, 1600–1860*. Cambridge: Cambridge University Press.

Heniger, J. (1986). *Hendrik Adriaan van Reede tot Drakenstein 1636–1691 and Hortus Malabaricus: A Contribution to the History of Dutch Colonial Botany.* Rotterdam: A.A. Balkema.

Hesselink, Liesbeth (2010). "Crossing Colonial and Medical Boundaries: Plural Medicine on Java," in Anne Digby, Waltraud Ernst, and Projit B. Mukharji (eds.), *Crossing Colonial Historiographies: Histories of Colonial and Indigenous Medicines in Transnational Perspective.* Newcastle upon Tyne: Cambridge Scholars Publishing, pp. 115–142.

Kloppenburg-Versteegh, J. (1934 [1911]). *Wenken en Raadgevingen Betreffende het Gebruik van Indische Planten, Vruchten Enz.* [Guidance and Advice Regarding the Use of Indies Plants, Fruits, etc.], 3rd ed., 2 vols. Semarang: G.C.T. van Dorp.

Latour, Bruno (1987). *Science in Action: How to Follow Scientists and Engineers through Society.* Cambridge: Harvard University Press.

Lyon, Margot L., "Jamu for the Ills of Modernity," *Inside Indonesia*, (2003) 75: 14–15.

Macknight, C.C. (1976). *The Voyage to Marege: Macassan Trepangers in Northern Australia.* Melbourne, VIC: Melbourne University Press.

Ong, Walter (1982). *Orality and Literacy: The Technologization of the Word.* London: Methuen.

Orta, Garcia da (1563). *Colóquios Dos Simples E Drogas He Cousas Medicinais Da Índia* [*Conversations on the Simples, Drugs and Medicinal Substances of India*]. Goa: Ioannes de Endem.

Pols, Hans. "The Triumph of Jamu," *Inside Indonesia* (2010) 100: http://www.insideindonesia.org/weekly-articles/the-triumph-of-jamu.

———— "European Botanists and Physicians, Indigenous Herbal Medicine in the Dutch East Indies, and Colonial Networks of Mediation," *East Asian Science, Technology, and Society: An International Journal* (2009) 3(2–3): 173–208.

Raj, Kapil (2008). *Relocating Modern Science: Circulation and the Construction of Knowledge in South Asia and Europe, 1650–1900.* Basingstoke: Palgrave MacMillan.

Reid, Anthony (1993). *Southeast Asia in the Age of Commerce, 1450–1680, Vol. 1: The Lands Below the Winds.* New Haven, CT: Yale University Press.

———— (1998). *Southeast Asia in the Age of Commerce, 1450–1680, Vol. 11: Expansion and Crisis.* New Haven, CT: Yale University Press.

Roberts, Lissa. "Situating Science in Global History: Local Exchanges and Networks of Circulation," *Itinerario* (2009) 33(1): 9–30.

Rumphius, Georgius Everhardus, and E.M. Beekman (1999 [1705]). *The Ambonese Curiosity Cabinet.* New Haven, CT: Yale University Press.

Rumphius, Georgius Everhardus (2011 [1741]). *The Ambonese Herbal, Volume 1: Introduction and Book 1: Containing All Sorts of Trees, That Bear Edible Fruits, and*

Are Husbanded by People. Translated by E.M. Beekman, 6 vols., vol. 1. New Haven, CT: Yale University Press.

———— (2011 [1741]). *The Ambonese Herbal, Volume 2: Book II: Containing the Aromatic Trees: Being Those That Have Aromatic Fruits, Barks or Redolent Wood; Book III: Containing the Wild Trees That Provide Timber.* Translated by E.M. Beekman, 6 vols., vol. 2. New Haven, CT: Yale University Press.

———— (2011 [1741]). *The Ambonese Herbal, Volume 3: Book V: Dealing with the Remaining Wild Trees in No Particular Order; Book VI: Concerning Shrubs, Domesticall and Wild; the Forest Ropes and Creeping Shrubs.* Translated by E.M. Beekman, 6 vols., vol. 3. New Haven, CT: Yale University Press.

———— (2011 [1741]). *The Ambonese Herbal, Volume 5: Book XII: Concerning the Little Sea Trees, and Stony Sea Growths, which Resemble Plants; Auctuarium, or Augmentation of the Ambonese Herbal.* Translated by E.M. Beekman, 6 vols., vol. 5. New Haven, CT: Yale University Press.

———— (2011 [1741]). *The Ambonese Herbal, Volume 4: Book VIII: Containing Potherbs Used for Food, Medicine, and Sport; Book IX: Concerning Bindweeds, as Well as Twining and Creeping Plants.* Translated by E.M. Beekman, 6 vols., vol. 4. New Haven, CT: Yale University Press.

———— (2011 [1741]). *The Ambonese Herbal, Volume 6: Species List and Indexes for Volumes 1–5.* Translated by E.M. Beekman, 6 vols., vol. 6. New Haven, CT: Yale University Press.

Schaffer, Simon, Lissa Roberts, Kapil Raj, and James Delbourgo (eds.) (2010). *The Brokered World: Go-Betweens and Global Intelligence, 1770–1820.* Sagamore Beach, MA: Science History Publications.

Schivelbusch, Wolfgang (1992). *Tastes of Paradise: A Social History of Spices, Stimulants, and Intoxicants.* Translated by David Jacobson. New York: Pantheon.

Smits, D. de Visser (1928). "Geneeskunde en Plantkunde [Medicine and Botany]," in *Handelingen van het Vijfde Nederlandsch-Indisch Natuurwetenschappelijk Congres, Soerabaja.* Batavia: Kolff, pp. 71–86.

Taylor, Jean Gelman (1983). *The Social World of Batavia: European and Eurasian in Dutch Asia.* Madison, WI: University of Wisconsin Press.

Van Blokland, Njonja (1899). *Doekoen Djawa: Oetawa Kitab dari Roepa-Roepa Obat njang Terpake di Tanah Djawa* [*Javanese Dukun or Book with Various Kinds of Medicine in Use on Java*]. Batavia: Albrecht & Co.

Van der Burg, C.L. (1884). "Land, Klimaat en Bewoners; Hygiëne: De Uitoefening van de Geneeskundige Praktijk" ["Land, Climate and Inhabitants; Hygiene: Engaging in Medical Practice"], in *De Geneesheer in Nederlandsch Indië* [*The Physician in the Dutch East Indies*], 2nd, improved and extended edition, vol. 1. 's Gravenhage: Martinus Nijhoff, pp. 384–398.

Van Gent-Detelle, Njonja E. (1880 [1875]). *Boekoe Obat-Obat voor [sic] Orang Toewa dan Anak-Anak [Medicine Boek for Adults and Children]*. Djocjacarta: Buning.

Van Hien, H.A. (1924). *Het Javaansch Receptenboek afkomstig van Soerakarta, bevattende 797 Recepten, voor Genezing van Ziekten van den Volwassen Mensch, het Kind, het Viervoetig Dier, den Vogel, en het Hoen, met vermelding tevens van de Meest Bekende Javaansche Vergiften, Geheime Middelen en van de Latijnsche Namen der in dit Boek vermelde Planten [The Javanese Book of Prescriptions from Surakarta, Containing 797 Recipes for Curing Diseases in Adults, Children, Animals and Birds, as well as Mentioning the Most Famous Javanese Poisons, Secret Means, and the Latin Names of the Plants Listed in this Book]*. Weltevreden: Visser & Co.

———— (1986). *De Javaansche Geestenwereld en de Betrekking die tusschen de Geesten en de Zinnelijke Wereld Bestaan, Verduidelijkt door Petanga's of Tellingen bij de Javanen in Gebruik*, The Javanese Spirit World and the Relationship that Exist between Spirits and the Sensual World, Clarified by *Petangas* or Narratives Common among the Javanese, 4 vols. Semarang: G.C.T. van Dorp.

Vickers, Adrian (2005). *A History of Modern Indonesia*. Cambridge: Cambridge University Press.

Wiener, Margaret J. (1995). *Visible and Invisible Realms: Power, Magic and Colonial Conquest in Bali*. Chicago, IL: University of Chicago Press.

———— (2003). "Hidden Forces: Colonialism and the Politics of Magic in the Netherlands Indies," in B. Meyer and P. Pels (eds.) *Magic and Modernity*. Stanford: Stanford University Press, pp. 129–58.

———— "Dangerous Liaisons and Other Tales from the Twilight Zone: Sex, Race, and Sorcery in Colonial Java," *Comparative Studies in Society and History* (2007) 49(3): 495–526.

CHAPTER 10

From Zero to Infinity: The Indian Legacy of the Bright Dark Ages

George Gheverghese Joseph

1   Introduction

A few years ago on a British television programme I was asked: "Why did zero originate in India?" Fortunately, I was allowed enough time to develop an answer without assuming, as most television programmes do today, that the audience watching have the attention span of a grasshopper. Trying to gather my thoughts, I resorted to the familiar ploy of taking refuge in definitions. If zero merely signified a magnitude or a direction separator (i.e. separating those above the zero level from those below the zero level), the Egyptian zero, *nfr*, dating back at least four thousand years, amply served these purposes. If zero was merely a place-holder symbol, indicating the absence of a magnitude at a specified place position (such as, for example, the zero in 101 indicates the absence of any "tens" in one hundred and one), then such a zero was already present in the Babylonian number system long before the first recorded occurrence of the Indian zero. If zero was represented by just an empty space within a well-defined positional number system, such a zero was present in Chinese mathematics a few centuries before the Indian zero. The absence of a symbol for zero did not prevent the Chinese numerals from being properly integrated into an efficient computational tool that could even handle solutions of higher degree order equations involving fractions. However, the Indian zero alluded to in the question was a multi-faceted mathematical object: a symbol, a number, a magnitude, a direction separator and a place-holder, all in one operating within a fully established positional number system. Such a discovery of zero occurred only twice in history – the Indian zero which is now the universal zero and the Mayan zero which occurred in splendid isolation in Central America around the beginning of Common Era.[1]

---

1   It is interesting in this context to recognise that a place value system can exist in the absence of a symbol for zero. The Babylonian and the Chinese number systems were good examples. But the zero symbol as part of a system of numerals could never have come into being without a place value system. In neither the Egyptian, Greek nor Aztec cultures was there a place

To understand the first appearance of the Indian zero, it is necessary to examine it within the social context in which it occurred. The dissemination of the Indian zero as part and parcel of Indian numerals is one of the more remarkable episodes in the history of mathematics. But what is rarely recognised is that this transmission occurred through a number of cultural and linguistic filters that may have inhibited a clearer understanding of the concept of zero and the arithmetic of the operations associated with it. Because of the widespread difficulties with the concept of zero when it came to be adopted outside India, there has grown over time a series of intellectual avoidance mechanisms which have had far-reaching pedagogical implications. And these include the general absence of any discussion on the topic of 'calculating with zero' (*shunya-ganita*), emphasized in a number of Indian texts on mathematics from the time of Brahmagupta (b. 598) onwards.

2   The History of Zero: The Indian Dimension

The word 'zero' comes from the Arabic *al-sifr. Sifr* in turn is a transliteration of the Sanskrit word *shunya* meaning 'void' or 'empty'. Introduced into Europe during the Italian Renaissance in the 12th century by Leonardo Fibonacci (and by his near contemporary Jordanus de Nemore a lesser known mathematician) as "cifra" from which emerged the present 'cipher'. In French, it became "chiffre", and in German "ziffer", both of which mean zero.

The ancient Egyptians never used a zero symbol in writing their numerals. Instead they had a zero used as a value or magnitude. A bookkeeper's record from the 13th dynasty (about 1700 BCE) shows a monthly balance sheet for items received and disbursed by the royal court during its travels. On subtracting total disbursements from total income, a zero remainder was left in several columns. This zero remainder was represented by the hieroglyph, *nfr*, which also means 'beautiful', or 'complete' in ancient Egyptian. The same *nfr* symbol also labelled a zero reference point for a system of integers used on construction guidelines at Egyptian tombs and pyramids. These massive stone structures required deep foundations and careful levelling of the courses of stone. A vertical number line labelled the horizontal levelling lines that guided construction at different levels. One of these horizontal lines, often at pavement level, was used as a reference and was labelled *nfr* or zero. Horizontal levelling lines were spaced 1 cubit apart. Those above the zero level were

---

value system. A zero as a number in any of these systems would in any case have been superfluous. For a discussion of the Mayan zero, see Joseph (2011), pp. 66–72.

labelled as 1 cubit above *nfr*, 2 cubits above *nfr* and so on. Those below the zero level were labelled 1 cubit, 2 cubits, 3 cubits, and so forth, below *nfr*. Here zero was used as a reference for directed or signed numbers.

It is quite extraordinary that the Mesopotamian culture, more or less contemporaneous to the Egyptian culture and which had developed a full positional value number system on base 60 did not use zero as a number. A symbol for zero as a place-holder appeared late in the Mesopotamian culture. The early Greeks, who were the intellectual inheritors of Egyptian mathematics and science emphasised geometry to the exclusion of everything else. They did not seem interested in perfecting their number notation system. They simply had no use for zero. In any case, they were not greatly interested in arithmetic, claiming that arithmetic should only be taught in democracies for it dealt with relations of equality. On the other hand, geometry was the natural study for oligarchies for it demonstrated proportions within inequality.

In India, the zero as a concept probably predated zero as a number by hundreds of years. The Sanskrit word for zero, *shunya*, meant "void or empty". The word is probably derived from *shuna* which is the past participle of *svi*, "to grow". In one of the early Vedas, *Rgveda*, occurs another meaning: the sense of "lack or deficiency". It is possible that the two different words, were fused to give "*shunya*" a single sense of "absence or emptiness" with the potential for growth. Hence, its derivative, *Shunyata*, described the Buddhist doctrine of Emptiness, being the spiritual practice of emptying the mind of all impressions. This was a course of action prescribed in a wide range of creative endeavours. For example, the practice of *Shunyata* was recommended in writing poetry, composing a piece of music, in producing a painting or any other activity that come out of the mind of the artist. An architect was advised in the traditional manuals of architecture (the *Silpas*) that designing a building involved the organisation of empty space, for it is not the walls which make a building but the empty spaces created by the walls. The whole process of creation is vividly described in the following verse from an 8th century Tantric Buddhist (*Hevjra Tantra*) text:

> First the realisation of the void (*shunya*),
> Second the seed in which all is concentrated,
> Third the physical manifestation,
> Fourth one should implant the syllable.

The implication here is that creativity in any field involves the capacity to realize the void and represent it. The mathematical correspondence was soon established. "Just as emptiness of space is a necessary condition for the appearance of any object, the the number zero being no number at all is the condition for the existence of all numbers." (Datta, 1927: 168)

A discussion of the mathematics of the *shunya* involves three related issues: (i) the concept of the *shunya* within a place-value system, (ii) the symbols used for *shunya*, and (iii) the mathematical operations with the *shunya*. In what follows we will use material from early texts as illustrations to illuminate these issues.

It was early recognised that since the *shunya* denoted notational place (place holder) as well as the "void" or absence of numerical value in a particular notational place, all numerical quantities, however great, could be represented with just ten symbols. This is noted in the twelfth century text *Manasollasa* or 'The Refresher of the Mind':

> Basically, there are only nine digits, starting from 'one' and going to 'nine'. By the adding of zeros these are raised successively to tens, hundreds and beyond.

The same point is made in a commentary on Patanjali's *Yogasutra* that appeared in the seventh century by the following analogy:

> Just as the same sign is called a hundred in the "hundreds" place, ten in the "tens" place and one in the "units" place, so is one and the same woman referred to (differently) as mother, daughter or sister.

The earliest mention of a symbol for zero occurs in the *Chandahsutra* of Pingala (fl. 2nd century BCE) which discusses a method for calculating the number of arrangements of long and short syllables in a metre containing a certain number of syllables (i.e. the number of combinations of two items from a total of n items, repetitions being allowed). The symbol for *shunya* began as a dot (*bindu*), found in inscriptions in India, Cambodia and Sumatra around the seventh and eighth century, which then became a circle (*chidra* or *randra* meaning a hole at a later time). The association between the concept of zero and its symbol was already well-established by the early centuries of the common era as the following quotation from a romantic tale testifies:

> The stars shone forth like zero dots (*shunya-bindu*) scattered in the sky as if on the blue rug, (on which) the Creator reckoned the total with a bit of the moon for chalk. (Vasavadatta, c. 400; quoted in Datta and Singh 1962: 76)

Sanskrit texts on mathematics and astronomy from the time of Brahmagupta usually contain a section called *shunya-ganita* or computation with zero. While the discussion in the arithmetical texts (*patiganita*) is limited only to the addition, subtraction and multiplication with zero, the treatment in algebra texts (*bijaganita*) covered such questions as the effect of zero on the positive and negative signs, division with zero, and more particularly the relation between zero and infinity (*ananta*).

Take as an example, Brahmagupta's seventh century text *Brahmasphuta-siddhanta*. In it he treats the zero as a separate entity from the positive (*dhana*) and negative (*rhna*) quantities, implying that *shunya* is neither positive nor negative but forms the boundary line between the two kinds, being the sum of two equal but opposite quantities. A negative number subtracted from zero is positive, a positive number subtracted from zero is negative, zero subtracted from a negative number is negative, zero subtracted from a positive number is positive, zero subtracted from zero is zero. A number, whether positive or negative, remained unchanged when zero is added to or subtracted from it. In multiplication with zero, the product is zero. A zero divided by zero or by any other number becomes zero.

But when a non-zero number is divided by zero, the answer is an undefined quantity. Bhaskara in the 12th century explained this conundrum thus:

> A quantity divided by zero becomes a fraction the denominator of which is zero. This fraction is sometimes termed an infinite quantity. In this quantity consisting of that which has zero for its divisor, there is no alteration, though many (quantities) may be inserted or extracted; as no change takes place in the infinite and immutable God when worlds are created or destroyed, though numerous orders of beings are absorbed or put forth. [Bhaskara II's *Bijaganita*, Section 2, Verse 11]

This could be taken to imply that any number divided by zero is infinity. Also, Bhaskara went on to suggest that zero multiplied by infinity is any number, and hence the implication that all numbers are equal, which is obviously not correct. But Bhaskara did guess correctly that the square of zero is zero, as is the square root.

3   The Spread of the Indian Numerals (and Zero)

The oldest known text to mention zero is a Jaina text entitled the *Lokavibhaga* ("The Parts of the Universe"), which has been dated to the fifth century BCE. The earliest inscription in India of a recognisable antecedent of our numeral system is found in an inscription from Gwalior dated 'Samvat 933' (876 CE). The spread of these numerals westwards is a fascinating story. The Islamic world was the venue for this drama. Indian numerals probably arrived at Baghdad in the year 773 with the diplomatic mission from Sindh to the court of Caliph al-Mansur. Around 820 al-Khwarizmi wrote his famous *Arithmetic*, the first Arabic text to deal with the new numerals. The book contains a detailed exposition of the representation of numbers and arithmetical operations using Indian numerals. Al-Khwarizmi was at pains to point out the usefulness of a place-value system incorporating zero, particularly for writing large numbers. Texts on Indian reckoning continued to be written in the Islamic world and by the end of the eleventh century, this method of representation and computation became widespread from the borders of Central Asia to the southern reaches of the Islamic world in North Africa and Egypt.

Zero reached Europe in the twelfth century when Adelard of Bath translated al-Khwarizmi's works into Latin. Fibonacci was one of the main mathematicians who introduced the concept of zero to Europe. In his treatise *Liber Abaci*, published in 1202, he described the nine Indian symbols together with the symbol 'o' for zero. Significantly, Fibonacci only used the word 'numbers' for 1 through to 9, but the word 'sign o' for zero. The Europeans were at first resistant to this system of numeration, being attached to the old Roman numeral system, but their eventual adoption of this system arguably led to the scientific revolution that began to sweep Europe beginning around the middle of the second millennium. However, it was not until the 17th century that zero found widespread acceptance.

From the time of Han (206 BCE to 220 CE), Chinese scholars used a place-value system called the *suan zi* ('calculation using rods') where different configurations of horizontal and vertical lines denoted the nine numerals and zero was represented by an empty space. The notion and representation of zero with a symbol helps one to recognize the point of departure of the Indian mathematicians vis-à-vis their Chinese counterparts. It should be noted that for the Chinese the absence of zeros posed little difficulty in representing numbers like 1,270,000 by either using characters of their ordinary counting system (a non-positional system that did not require the use of a zero) or simply by empty spaces. It was only after the eighth century, and due to the Indian influence through the Buddhist missionaries that Chinese mathematicians were first introduced the use of zero in the form of a little circle or dot. In any case, by the 12th and 13th centuries, mathematicians Chin Chiu-Shao and Chu Shih-Chieh were familiar with same symbol o within a decimal-based system.

There is evidence of the earlier use of the Indian system of numeration in South East Asia in areas covered by present-day countries such as Malaysia, Cambodia, Vietnam and Indonesia, all of whom were under the cultural influence of India.[2] There, three inscriptions have been found bearing dates in the Saka era, which began in 78 CE. A Malay inscription at Palembang in Sumatra from 684 CE shows the zeroes in 60 and 606 Saka as '0', and a Khmer inscription at Sambor in Cambodia from 683 CE gives the zero in Saka year as a dot (.). An inscription at Ponagar, Champa (now southern Vietnam) from 813 CE represents the year as 735 in Indian notation.[3]

4   The Concept of Infinity: The Indian Contribution

The *Isha Upanishad* of the *Yajurveda* (c. 4th to 3rd century BCE) states: "If you remove a part from infinity or add a part to infinity, still what remains is infinity". The essence of this statement is that the infinite cannot be measured arithmetically and that the infinite can be represented (and manifests) in infinite ways. The idea of mathematical infinity evolved largely due to the Jaina's cosmological ideas. In Jaina cosmology, time is thought of as eternal and without form. The world is infinite, it was never created and has always existed. Space pervades everything and is without form. All the objects of the universe exist in space which is divided into the space of the universe and the space of the non-universe. There is a central region of the universe in which all living beings, including men, animals, gods and devils, live. Above this central region is the upper world which is itself divided into two parts. Below the central region is the lower world which is divided into seven tiers. This led to the work described in *Tiloyapannatti* by Yativrsabha. A circle is divided by parallel lines into regions of prescribed widths. The lengths of the boundary chords and the areas of the regions are given, based on a set of rules. The first group, the enumerable numbers, consisted of all the numbers from 2

---

2 We may infer that the Indian computational methods would have been known to the Javanese astronomers. H. Kern and J.G. de Casparis have identified inscriptions of Java 700–1500 that refer to specific Sanskrit texts. Indo-Javanese astronomers and astrologers have used the Indian concept of *muhurta* (i.e. an 'hour' each of 48 minutes) to compute auspicious times for various undertakings. The old Javanese nomenclature for *muhurta* hours appear in a rare Sanskrit astrological text *Atharvanajyotisa*. Terms such as *mandala* and *parvesa* have been used in earthquake divinations and eclipse computations respectively. Both concepts originate either directly or indirectly from the popular Sanskrit divination text, Varahamihira's *Brhatsamhita* (6th century CE). Although it has not been possible to determine the specific Sanskrit *Siddhanta* texts used by the Indo-Javanese astronomers, computational verification showed that they closely followed the theories of *Siddhanta* texts. For further details, see Salomon (1998).

3 For a more detailed discussion, see Joseph (2011), p. 339.

(*1* was ignored) to the highest. An idea of the 'highest' number is given in the following extract from the *Anuyoga Dwara Sutra*, from around the beginning of the contemporary era:

> Consider a trough whose diameter is that of the Earth (100000 *yojanna*) and whose circumference is 316 227 *yojanna*. Fill it up with white mustard seeds counting one after another. Similarly fill up with mustard seeds other troughs of the sizes of the various lands and seas. Still the highest enumerable number has not been attained. (1 *yojanna* is about 10 kilometers.)

But if and when this number, call it *N*, is attained, infinity is reached via the following sequence of operations:

$$N+1, N+2 \ldots \ldots \ldots \ldots \ldots, \quad (N+1)^2 - 1$$
$$(N+1)^2, (N+2)^2 \ldots \ldots \ldots, \quad (N+1)^4 - 1$$
$$(N+1)^4, (N+2)^4 \ldots \ldots \ldots, \quad (N+1)^8 - 1 \quad \text{and so on}$$

This cosmology has strongly influenced Jaina mathematics in many ways and has been a motivating factor in the development of mathematical ideas of the infinite which were not considered again until the time of Cantor.[4]

As can be seen, enormous numbers were thought of from a very early time period in India and it is interesting to note the extent to which these numbers became part of popular culture. The handling of such large numbers (at first in the oral tradition) probably spurred the development of the decimal number system with place value and zero. Several stages involving symbolization, concept of place value and zero were crossed before the final construction of the number system. A detailed discussion of this is beyond the scope of this paper. However, one significant aspect in terms of pedagogy is that in the naming of

---

4   Like the Vedic mathematicians, the Jains had an interest in the enumeration of very large numbers which was intimately tied up with their philosophy of time and space. They devised a measure of time, called a *shirsa prahelika*, which equalled $756 \times 10^{11} \times (8,400,000)^{28}$ days! Other examples of the Jaina fascination with very large numbers are two definitions: a *rajju* is the distance travelled by a god in six months if he covers a hundred thousand *yojanna* (approximately a million kilometres) with each blink of his eye; a *palya* is the time it will take to empty a cubic vessel of side one *yojanna* filled with the wool of new-born lambs if one strand is removed every century.

large numbers, the multiplicative structure (and particularly the exponential) was present from early Vedic times.[5]

5   The Passage to Infinity: The Kerala Episode

Two powerful tools contributed to the creation of modern mathematics in the seventeenth century: the discovery of the general algorithms of calculus and the development and application of infinite series techniques. These two streams of discovery reinforced each other in their simultaneous development since each served to extend the range and application of the other.

Existing literature would lead us to believe that the methods of the calculus were invented independently by Newton and Leibniz in the middle of the seventeenth century, building on the works of their European predecessors such as Fermat, Taylor, Gregory, Pascal and Bernoulli (Edwards, 1979) in the preceding half century. What appears to be less well known is that certain fundamental elements of the calculus including infinite series derivations for π and certain trigonometric functions such as $\sin x$, $\cos x$ and $\tan^{-1} x$ (the so-called Gregory series) were already known in India to Kerala mathematicians more than 250 years earlier.

The primary mathematical motivation for the Kerala work on infinite series arose from a recognition of the impossibility of arriving at an exact value for the circumference of a circle given the diameter (i.e., the incommensurability of π). This recognition was prompted by a passage in Aryabhata's mathematical text, *Aryabhatiya*, written in 499 CE. Verse 10 of the section on *Ganita* in this work reads:

> Add 4 to 100, multiply by 8, and add 62,000. The result is *approximately* the circumference of a circle whose diameter is 20,000.

It was the word "approximately" that gave succeeding mathematicians food for thought.

---

5   In the Jaina work on the theory of sets, two basic types of transfinite numbers (i.e. the cardinal numbers of infinite sets) are distinguished. On both physical and ontological grounds, a distinction is made between *asmkhyata* and *ananta*, between respectively rigidly bounded and loosely bounded infinities. With this distinction, the way was open for the Jains to develop a detailed classification of transfinite numbers and mathematical operations for handling transfinite numbers of different kinds. However, unsurprisingly they did not do so given their limited technical and symbolic compass.

It led Nilakantha, one of the major Kerala mathemtaicians who wrote a commentary, *Aryabhatiyabhasya*, on Aryabhatiya's study, nearly a millennium later to explain why only an approximate value of the circumference may be obtained for a given diameter in the following terms:

> Why is only the approximate value (of circumference) given here? Let me explain. Because the real value cannot be obtained. If the diameter can be measured without a remainder, the circumference measured by the same unit (of measurement) will leave a remainder. Similarly, the unit which measures the circumference without a remainder will leave a remainder when used for measuring the diameter. Hence, the two measured by the same unit will never be without a remainder. Though we try very hard we can reduce the remainder to a small quantity but never achieve the state of 'remainderlessness'. This is the problem.

This in turn motivated mathematicians of the Kerala school to try to find the length of an arc by approximating it to a straight line using a method known as direct rectification of an arc of a circle. The method involves summation of very small arc segments and reducing the resulting sum to an integral. The tangent is divided up into equal segments while at the same time forcing a subdivision of the arc into *unequal* parts. This is required since the method involves the summation of a large number of very small arc segments, traditionally achieved in European and Arab mathematics by the 'method of exhaustion', where there was a sub-division of an arc into *equal* parts. The adoption of this "infinite series" technique rather than the "method of exhaustion" for implicitly calculating π was not through ignorance of the latter in Indian mathematics. But, as Jyesthadeva, a Kerala mathematician of the sixteenth century, points out in the *Yuktibhasa* – a work that summarises the achievements of the Kerala School over the previous two centuries – the method of rectification avoids tedious and time-consuming root-extractions.

6   Conclusion: Zero and Mathphobia

The spread of the Indian zero had to go through a number of cultural and linguistic filtering processes, the imperfect nature of which generates discomfort even today. Underlying such uneasiness is both a conceptual fuzziness regarding zero and a lack of confidence in the manipulation of mathematical expressions where the notions of zero or infinity present themselves. A story told of youthful Srinivas Ramanujan, the famous twentieth century Indian mathematician, illustrates this point well. In an elementary mathematics class the teacher was explaining the concept of division (or 'sharing') through examples. If three bananas were shared between three children, each child would get one banana. And similarly, the share would be one banana if four bananas were divided among four children, five bananas among five children and so on. And when the teacher generalised this idea of sharing out x bananas among x boys, Ramanujan piped up with the question: "If x equalled zero, would each child then get one banana?" There is no record of the teacher's reply.

Ask a mathematician whether zero is an even or an odd number? The answer would be: If you define evenness or oddness on the integers (either positive or all), then zero should be taken to be even; but if you define evenness and oddness on the natural numbers, then zero would be neither. This is because we apply concepts such as "even" only to "natural numbers," in connection with primes and factoring, where by "natural numbers" one means positive integers and so excludes zero. However, those who work in foundations of mathematics consider zero a natural number, and for them the integers are whole numbers. From that point of view, the question whether zero is even just does not arise, except by extension. One may say that zero is neither even nor odd. Because you can pick an even number and divide it in groups, take, e.g., 2, which can be divided in two groups of "1", and 4 can be divided in two groups of "2". But can you divide zero? This was basically Ramanujan's question. And the difficulty is basically caused by the fact that the concepts of even and oddness predated zero and the negative integers.

On the question of division by zero there seems to be confusion found even among seasoned practitioners. It is not uncommon even today to came across, for example, the division of 2 by zero in performing a column ratio test as part of constructing a Simplex Tableau (an algorithmic engine for solving linear programmes) the conclusion, $2 \div 0 =$ infinity. However the logic of this conclusion is not followed through in the simplex calculation based on this result. Neither does one detect the erroneous nature of this reasoning by asking the obvious question: Which number, when multiplied by zero, gives you 2? *Infinity?* But is infinity really a number?

Another commonly given answer is: *Undefined.* Is this correct?[6] If we try to find an answer by looking for it in a pocket calculator, we get an error notification, suggesting that one cannot ever meaningfully divide by zero. For if

---

6 The problem with this argument is easily seen if we represent $2/0 = x$. The question then becomes: "What is the value of x?" It could be any number. One number cannot be equal to so many different numbers. Therefore, teaching students that $2/0 =$ Any Number (AN) is equivalent to saying that AN x 0 = 0 which clearly contradicts the statement we began with.

one allows division by zero, then one enters a topsy-turvy world where $1 = 2$.[7] Clearly zero is a number but not similar to other numbers when the arithmetical operation of division is involved.

Given these difficulties, one popular response is to avoid the imagery associated with this seemingly difficult concept by using euphemisms such as nought, 'o', nothing, etc. In reciting a telephone number, a postal zip code or a street number or any of a variety of other numerical codes, we often try to avoid the use of the digit name "zero". All the *other* digits are correctly enunciated with this one being the curious exception. In tennis scores, zero is called "love". The French called it "l'oeuf," which was corrupted to 'love'. Zero is placed as the last number on a computer keyboard after all other digits. It appears at the bottom of the keypad on a telephone. There is even resistance in multi-storey buildings to label the ground-level as the 'o' level. However, it is interesting where such a resistance is absent, as for example in the case of certain buildings in Eastern Europe (and also in South America) floors are labelled as –1, 0, 1, 2, 3. . . . , with – 1 representing the basement. This practice is instructive. It indicates that in the absence of a concept of zero there could have been only positive numerals. The incorporation of zero in mathematics opened up the new dimension of negative numerals. Incidentally, it is precisely because negative numbers seem to have first appeared in Chinese mathematics, that the historian of mathematics, Lam Lay Yong, argued that the zero was invented by the Chinese – a 'symbol-less' zero in this case. (Lam & Ang, 1992)

It is as though the name 'zero' itself invokes a kind of anxiety perhaps associated with "nothingness", a kind of emptiness which humankind finds uncomfortable and prefer to avoid confronting. As with all such anxiety-provoking ideas, some other imagery is substituted which provides a veneer to mask the disquieting emotional undertones of the discomforting idea.[8] Yet in the culture where zero was first used as a number within a well-developed positional value system, the concept was devoid of any negative overtones. This difference merits further investigation.

---

7   To show that $1 = 2$, for any finite a:

  $(a).(a) - a.a = a^2 - a^2 = > a(a\text{-}a) = (a\text{-}a)(a\text{+}a)$

  Dividing both sides by $(a\text{-}a)$ gives

  $a = 2a = > 1 = 2$ Behold!

8   Sometimes, this anxiety could lead to a uncritical acceptance of numerology. An example associated with zero is found in the Indian Chief's Tecumesch's Curse which has allegedly killed every U.S. President before the end of their term in office, if they were elected in a year that ended with 0. The last victim of the curse was John F. Kennedy who was elected in 1960. See Henderson and Nugent (1980).

Bibliography

Datta, B. "Early history of the Arithmetic of Zero and Infinity in India," *Bulletin of the Calcutta Mathematical Society* (1927) 18: 165–176.

Datta, B. and A.N. Singh (1962). *History of Hindu Mathematics: A Source Book*, Part I. Bombay: Asia Publishing House.

Henderson, R. and T. Nugent (1980). "The Zero Curse: More Than Just a Coincidence?" November 2, 1980, in *Syracuse Herald-American*, p. C3. (reprinted from the *Baltimore Sun*).

Joseph, G.G. (2011). *The Crest of the Peacock: Non-European Roots of Mathematics*. Princeton: Princeton University Press.

Lam Lay Yong and Ang Tian Se (1992). *Fleeting Footsteps: Tracing the Concept of Arithmetic and Algebra in Ancient China*. Singapore: World Scientific.

Salomon, R. (1998). *Indian Epigraphy: A Guide to the Study of Inscriptions in Sanskrit, Prakrit and other Indo-Aryan Languages*. New York: Oxford University Press.

CHAPTER 11

The Needham Question and Southeast Asia: Comparative and Connective Perspectives

Arun Bala

1   Needham's Grand Question: Marginalizing Southeast Asia

There has been a renewed revival of interest in Needham's Grand Question, "Why did modern science develop in Europe but not in civilization X?" where X is a marker for any one of the world's historical civilizations. Part of the revival can be attributed to a new interest in connecting the Needham question with the Weber question "Why did the Industrial Revolution take place in Europe and not elsewhere?" Equally important are two recent attempts to answer the Needham question through a comparative study of civilizations – Floris Cohen's *How Modern Science Came into the World: Four Civilizations, One 17th-Century Breakthrough* and Toby Huff's *Intellectual Curiosity and the Scientific Revolution: A Global Perspective*. Both build upon and amplify earlier studies – Floris Cohen's *The Scientific Revolution: A Historiographical Inquiry* and Toby Huff's *The Rise of Early Modern Science: Islam, China and the West*. These studies can be seen as attempting to understand the rise of modern science in the West in the context of its failure to emerge in the most advanced civilizations that preceded it – especially the Chinese and the Islamic-Arabic. Nevertheless, the answers Cohen and Huff give are largely predicated on a comparative perspective that only marginally takes into account recent attempts to see the Scientific Revolution as shaped by a connective history in which the civilizations they compare to the West – the Chinese and Islamic – also played a role in contributing to the intellectual, technological and technical resources that made possible the rise of modern science in the West. This study examines how our responses to the Needham question get transformed when we address them in a connective perspective that does not necessarily exclude the comparative dimension.

The above studies by Cohen and Huff have mainly focused on comparisons, and sometimes connections, between the Chinese, Islamic, and to some extent Indian civilizations with that of the West to explain why modern science emerged in the latter. I would like to suggest that it would be illuminating to extend comparative studies to include cultures which have hitherto been ignored because they were deemed to have no significant premodern traditions of science. Indeed, in his earlier study *The Scientific Revolution: A Historiographical Inquiry*, Floris Cohen does devote his whole penultimate chapter to attempting an answer to the Needham Question by including such less advanced cultures although he dismisses them precisely on these grounds as having no possibility of producing modern science.

Cohen divides civilizations into four categories – those without science, those with some science, those with advanced science, and the one civilization that produced modern science. As examples of civilizations without science he lists the Roman Empire and Southeast Asia; India he treats as a civilization with some science; China and the Arabic world had advanced science prior to the modern era; but only the West produced modern science. Turning to Needham's question he sees it as unnecessary to ask why neither the Roman Empire nor Southeast Asia *did not* produce modern science – these cultures were simply not developed enough to do so.[1] He considers even India to be only a civilization with some science but not one sufficiently advanced to merit asking the question, why modern science did not emerge there. Nevertheless, he is prepared to concede that new discoveries may make us rethink the issue. According to Cohen only the Islamic and Chinese worlds had scientific traditions that were sufficiently developed to compel us to seriously investigate why they did not produce modern science.[2]

Cohen's answers to the Needham question in these two cases may be summed up briefly as follows. Although he concedes that the philosophy of organic materialism, upon which Chinese science was constructed, did facilitate the Chinese to go beyond primitive thought and make more technological advances than Europe before the Scientific Revolution, he also thinks that it

---

1   Interestingly, Said al-Andalusi in his 11th century text comparing the scientific contributions of various civilizations lists the Chinese as having made none, but the Romans as having made some. See especially Chapters 3 and 9, al-Andalusi (1991); also Low, Morris F. (ed.) (1999).

2   Cohen's view is questionable. Although from the perspective of modern science India may not appear to have been at the cutting edge of science, the perspective was quite different when Said al-Andalusi wrote in the eleventh century from Europe:

> To their credit, the Indians have made great strides in the study of numbers and of geometry. They have acquired immense information and reached the zenith in their knowledge of the movements of the stars [astronomy] and the secrets of the skies [astrology] as well as other mathematical studies. After all that, they have surpassed all other peoples in their knowledge of medical science and the strengths of various drugs, the characteristics of compounds, and the peculiarities of substances. (al-Andalusi 1991: 12)

failed to carry the potential for modern science that was inherent in Europe's
Aristotelian heritage. He writes:

> [T]here appear indeed to be at least two roads which lead out of primi-
> tive thought. Both of these were fully worth pursuing; neither possessed
> an apparent in-built superiority. Just as Su Sung's water-clock was prob-
> ably a more accurate timekeeper than its mechanical counterpart in
> medieval Europe, just so one may well, if one likes, call conceptions of
> nature in traditional China more advanced in several respects than natu-
> ral philosophy in Europe prior to the Scientific Revolution. Only, the one
> approach happened to possess greater developmental possibilities than
> the other. Organic materialism ultimately ran into a 'magnificent dead
> end', whereas somehow out of Aristotelianism and some other, scattered
> (and meanwhile enriched) remnants of the Greek legacy a kind of science
> could be forged that went on to conquer the world. (Cohen 1994: 475)

By the scattered and enriched remnants Cohen refers to the heritage of ancient
Greek atomism derived from Democritus and the heritage of Greek mathematics, especially geometry, as systematized by Euclid. Although Chinese organic
materialism appeared to have made great advances in science, and especially
technology, it ultimately was surpassed by the new tradition forged out of the
Greek heritage in the early modern era – a tradition that itself came to define
and constitute such an era.

Turning to the case of Arabic civilization he acknowledged that it had a
better opportunity since it inherited Aristotelian science and relevant associated 'remnants'. But Cohen thinks that if a modern scientist, such as Galileo,
had indeed appeared in the Islamic culture which inspired Arabic science,
he would in all likelihood not have prevailed as Galileo did in the West.
(Cohen 1994: 471) This can be expected because, unlike the case in Europe,
educated persons in Arabic culture treated philosophy, and especially natural
philosophy, as a foreign science. Moreover, Muslim scientists, in contrast to
their European counterparts, also identified at heart with the objections raised
against their activities by their religious communities because the separation
of state and religion, along the lines of the co-existence of Roman and canoni-
cal law, did not exist in Arabic civilization. But this, assumes Cohen, is a necessary precondition to pave the way for the sort of independent thinking that
made possible early modern science. (Cohen 1994: 417)

In his attempt to deal with the question of why the science of the Bright
Dark Ages in Asia did not lead to modern science, Cohen assumes throughout that something immanent within Europe – its science, philosophy, or

society – made this transformation possible. Moreover, civilizations are judged by him in terms of how advanced their premodern traditions of science were in order to explain why they did not tip over into modern science. Rome and Southeast Asia, and even possibly India, were never in the running because they did not have advanced science. China and the Arabic world had advanced science, but did not produce modern science for lack of the right mix of intellectual and socio-cultural factors.

In his book *Intellectual Curiosity and the Scientific Revolution*, the social historian, Toby Huff offers a different, albeit related, explanation for why modern science was mainly the single-handed achievement of the West. He explicitly states this as the central motive for his book.

> [T]he present narrative will show, whatever glories ancient China, India, or the Islamic Middle East may have enjoyed in the past, their contributions to the making of modern science were minor. (Huff 2011: ix)

He adds later his grounds for such a claim admitting that many people may find it discomforting:

> It would be pleasant to think that all the people of the world shared equally in the extraordinary advance of thought signified by the Scientific Revolution, yet the European contribution far exceeded that of all the other peoples and civilizations of the globe. In the context of today's multiculturalism, this statement will sound like a Eurocentric sentiment. But as this study shows, in vast areas of scientific enquiry, such as optics, the science of motion, human anatomy, microscopy, pneumatics, and electrical studies, there were no parallels to Europe's discoveries outside the West in the seventeenth century. (Huff 2011: 8)

Huff also rebuts the excuse that the East had not investigated these phenomena for lack of opportunity. By using the telescope to make his case, he argues that it was not lack of opportunity but a deficit of curiosity that was the problem in the East. Noting that the telescope reached the East shortly after its invention in Europe in 1608, he writes:

> [T]he arrival of the telescope in China, India, and the Ottoman Middle East did not produce a similar upsurge in scientific curiosity. We can now see that a comparative approach starting with a different field – say pneumatics or microscopy – would have produced the same result:

extraordinary advances in Europe but few or no parallel advances outside Europe. Nevertheless, the failure of the telescope to trigger an excit-
ing new burst of scientific creativity, especially in astronomy, around the
world serves best to highlight a deficit in scientific curiosity that seems
to have prevailed outside Europe from before the seventeenth century
all the way to the end of the twentieth century. That is an extraordinary
record of cultural disparity. (Huff 2011: 300)

Huff attributes the rise of intellectual curiosity in the West to the autonomy of
medieval universities and the tradition of Aristotelian philosophy and its interest in nature that became integrated into European education at the time of the
Scholastic Renaissance in the twelfth and thirteenth centuries which involved
the "extraordinary fusion of Greek philosophy, Roman law and Christian
theology" that gave European civilization a coherence that the Chinese, Indian
and Islamic civilizations lacked. It also promoted the notion of the medieval
university as a corporate entity with legal autonomy to run itself free from
external interference. (Huff 2011: 147–148) However, equally significant for Huff
is the change in the educational agenda of the universities. According to Huff:

> The Europeans refocused the curriculum of the universities on the three
> philosophies: natural philosophy, moral philosophy, and metaphysics.
> Then they placed at the center of this new curriculum the natural books
> of Aristotle.... Put differently, the Europeans institutionalized the study
> of the natural world by making it the central core of the university cur-
> riculum. (Huff 2011: 150–151)

Hence both Cohen and Huff assign a crucial role to Aristotelian natural philosophy in paving the way to modern science in Europe. Nevertheless, assign-
ing Aristotelian philosophy such a role is questionable as we will now proceed
to show.

2   Cohen, Huff and Aristotelian Philosophy

Let us begin by examining Cohen's argument that Chinese organic materialism
hindered the birth of modern science and Aristotelian philosophy facilitated
it when it became integrated into the European medieval Scholastic tradition.
This is a disputable view since Aristotelianism, like the Chinese tradition, also
embraces an organic conception of nature which modern science subverted

when it came to articulate its new mechanical philosophy of nature. Indeed there are remarkable similarities between Aristotelian philosophy and Chinese organic materialism that make the above argument suspect.[3]

Aristotelian organicism has a greater affinity to Chinese organic materialism than either have to modern science. Both are based on a theory of five elements – earth, fire, water, air and quintessence in the Aristotelian case; and earth, water, fire, wood and metal in the Chinese case. Similarly, both assume a theory of four qualities – hot, cold, wet and dry. Even the notion of action-at-a-distance, largely repudiated by the mechanical conceptions that inspired modern science, is taken for granted in the Greek and Chinese organic natural philosophies. Moreover, far from growing out of Aristotelian thought, early modern scientists defined themselves against Aristotelian ideas in what they characterized as 'a quarrel of the ancients and the moderns'.[4] Hence, if Aristotelian organicism did not preclude modern science from emerging in Europe, it is not clear why Chinese organic materialism should have constituted, as Cohen assumes, an insurmountable obstacle.

Cohen's case for why modern science did not emerge in Arabic civilization is also questionable. In the first three centuries of Arabic civilization Aristotle's philosophy combined with Plato's, in an Aristotelian Platonism, as a dominant movement. It only declined in the twelfth century after al-Ghazali articulated his devastating critique against Greek natural philosophy. By contrast, Aristotelian philosophy was rejected as pagan in Europe after the fall of Rome, only assimilated for a brief period following the scholastic age, before being subverted in the early modern era. Hence, Arabic civilization was inspired by Aristotelian philosophy and science for a longer period than its dominance in the West. It raises the question: Why did modern science not emerge out of the Aristotelian womb in Arabic civilization before faith reasserted itself following the trials and tribulations of the Mongol invasions? Cohen does not address this question.[5]

Moreover in Islamic civilization Aristotle was considered to be *The Philosopher*, and Aristotelian philosophy was a dominant trend in the Islamic

---

3   Even in the case of ethics it has been argued that there are close affinities between Aristotelian virtue ethics and Confucian ethics. See Yu (2007).

4   Boruchoff argues that the printing press, firearms and the nautical compass which made it possible for the moderns to communicate, exert power, and travel distances never imagined by the ancients were used to support the argument that the former were superior to the latter. See Boruchoff (2012) pp. 133–63.

5   Moreover Saliba (2007) argues that Arabic science, especially astronomy, continued to make significant progress even under Mongol rule and beyond.

big row. Too narrow

world for over three hundred years from the ninth to the eleventh centuries. This makes it reasonable to ask why modern science did not emerge in the Islamic world. Cohen's explanation for this – the Mongol invasions, religious inhibitions – are not credible since these developed more than five centuries after the rise of Islamic civilization whereas modern science emerged in Europe in less than four centuries after the Scholastic revival of Aristotelian philosophy.

Similarly Huff's explanation also curiously appeals to Aristotelian natural philosophy – both in the way it became integrated into the Scholastic tradition by promoting the autonomy of universities and the interest in natural philosophy it nurtured. I say curious because the Scientific Revolution began by overthrowing Aristotelian natural philosophy with its organic conception of nature, by a mechanical vision of the cosmos which developed in Royal Societies funded by trading companies outside the control of church-dominated universities.

Moreover, the objects of intellectual curiosity Huff so carefully documents in his study only serves the case I am presenting – they show that Europeans exhibited a remarkable curiosity, lacking in other civilizations, for discovering, collecting and articulating mechanical technologies.[6] What needs explanation is not the lack of curiosity of other cultures but the interest Europeans showed in certain kinds of technologies that others did not see as important. I would like to suggest that the type of inquisitiveness Europeans exhibited was designed to enable them, in the early modern era, to participate in the wide trading network of the Asian civilizations which Europe joined in the post-Vasco da Gama age. Europe needed mechanical knowledge (ideas, computing techniques, technologies) in a way not required by other major civilizations. First, Europe could only join the Pan-Asian trading network by navigating long distances. Second, Europe had to display superior maritime artillery and ballistics in order to keep its stake in the Asian waters. Finally, Europeans found that they had little to offer the Asian civilizations by way of trade except the bullion they mined from the Americas.

Meeting all of these challenges posed significant problems that can be seen as linked to the mechanical sciences. Long-distance navigation posed problems of celestial mechanics and the tides; maritime artillery posed problems of predicting precisely the free fall of bodies and trajectories of projectiles; and mining for silver and other precious metals posed problems of simple

money

---

6 This was to have portentous consequences for the future that only became clear after the knowledge of mechanical processes and the mechanical world view became consolidated in the industrial revolution of the late eighteenth century.

machines, atmospheric pressure, the pump and so on. The importance of such mechanical knowledge was early recognized by Francis Bacon:

> It is well to observe the force and virtue and consequences of discoveries. They are to be seen nowhere more conspicuously than in those three which were unknown to the ancients, and of which the origin, though recent, is obscure and inglorious: namely, printing, gunpowder, and the magnet. For these have changed the whole face and state of things throughout the world, the first in literature, the second in warfare, the third in navigation; whence have followed innumerable changes; insomuch that no empire, no sect, no star, seems to have exerted greater power and influence in human affairs than these three *mechanical discoveries*. [My emphasis] (quoted in Needham 1954: 19)

Thus, in the early stages of the post-Vasco da Gama era, which began in 1498, Europe began by collecting a vast body of mechanical techniques, technologies and empirical practices. Later the integration of these resources drove theoretical thinking by stages until nearly two hundred years later Newton's synthesis in the *Principia* united these different problems into one conceptual framework. Even here his seminal discovery was to bring together navigation, artillery and mining with his insight that a planet was really a projectile falling under gravity towards the sun endlessly, and that the atmosphere is a fluid compressed by gravity which affected the working of pumps in mines. His synthesis consolidated mechanical natural philosophy as an alternative vision of the cosmos to the various organic conceptions of the Bright Dark Ages.

Nevertheless there continues to remain a puzzle. The Chinese, as we know, were also the creators of great mechanical technologies even if they did not see them as the highest forms of knowledge – this they assigned to political and moral concerns. Why was it then that it was the Europeans who produced modern science and not the Chinese? This is a question we will take up in the next sections.

3   Strange Parallels: Southeast Asia and Northwest Europe

The Needham question becomes more puzzling when we think that both Northwestern Europe and Southeast Asia had remarkable parallels before the modern era. It is significant that during the Bright Dark Ages both Southeast Asia and Europe were being drawn into the global trading network centered

E L Jones

on Asia connecting the Islamic world in Westasia, the Hindu-Buddhist world in Southasia and the Confucian Chinese world in Eastasia. The time of the construction of Borobodur in Java in 800 CE was the time of the crowning of Charlemagne in Rome as Holy Roman Emperor; the time of the scholastic renaissance in Europe in the 12th–13th centuries was the time of the Khmer Empire that built Angkor Wat. Both Northwestern Europe, excluding the Iberian West under Islamic rule and Eastern Europe under Mongol domination, and Southeast Asia benefited by being peripheries that came to be connected to the dominant structures of civilization and commerce centered in Westasia, Southasia and Eastasia.

Moreover these connected histories are not completely accidental when we consider Victor Lieberman's attempts to locate Southeast Asian and European histories within the context of a larger Eurasian framework. Lieberman argues that mainland Southeast Asia showed long term trends that had parallels in Europe involving "the combination of accelerated political integration, firearms-based warfare, broader literacy, religious textuality, vernacular literatures, wider money use, and more complex international linkages." (Lieberman 2003: 79) Oddly he sees island Southeast Asia as having more in common with what he terms the Eurasian 'heartland' – China, the Middle East and India – because all of these were ruled by conquering elites from outside – Manchus, Turks, Persians, Dutch and Iberians – and therefore became politically fragmented in the early nineteenth century. (Lieberman 2003: 80) According to Lieberman the trajectories taken by mainland Southeast Asia and Europe appeared to be synchronized in a fashion largely governed by Eurasia-wide climatic, commercial, and military stimuli.

However, Lieberman's parallels exacerbate even further the puzzle of the Needham Question. We have seen Cohen argue that the Needham question cannot be relevant for Southeast Asia because it never had any advanced science in the first place. For the same reason Huff does not take Southeast Asia into his comparative analysis of civilizations to explain why modern science emerged in Europe. At the same time Cohen argues that the question is also not relevantly applicable to Roman civilization. Yet paradoxically modern science can be said to have emerged precisely in those territories that saw themselves as direct descendants of the Roman Empire. This is arguably why the birth of Europe as an identifiable cultural entity is often traced to the crowning of Charlemagne as Holy Roman Emperor in 800 CE. Even when Huff, and indeed Cohen, point to the separation of state and religion in Roman law as crucial for the emergence of modern science, they are tacitly acknowledging that this science emerged in a civilization that can be deemed in many ways to be the descendant of Rome. Hence it becomes reasonable to ask why Europe should have produced modern science when Southeast Asia is not even considered to be in the running.

4   Connective Histories: Rethinking Needham's Question for
         Southeast Asia

However, in recent years a new conception of the history of modern science has emerged which suggests that it is not necessarily the civilization with the most advanced science that should be considered the most likely to produce modern science. It is the outcome of the recognition that modern science did not grow organically out of the philosophical, scientific and social soil of Europe, but that Europe created modern science by drawing on ideas, technologies and institutions from across the world, and especially from the Arabic and Chinese traditions of science in the Bright Dark Ages.

The change was brought about by Needham's pioneering and monumental effort in the numerous volumes of *Science and Civilization in China* to document the technological and intellectual contributions that Chinese science from the Bright Dark Ages made to modern science. In a sense, Needham's achievement may be seen as continuing a process of development of historical understanding that widened awareness of the multiple sources of influence on modern science. In the seventeenth century the general view held by many of the pioneers of the Scientific Revolution was that modern science came to supplant ancient Greek science by simply displacing it as a congeries of errors.[7] But by the nineteenth century historians began to see Greek science as the pioneering influence that made possible modern science.[8] In effect a tunnel of history was created connecting modern science with ancient Greek science. The intervening period was seen as one of decline – the so-called Medieval Dark Ages. In the early twentieth century Duhem took historical understanding further by documenting how modern science profited from many philosophical and scientific ideas articulated in the scholastic period, so that the tunnel of history deepened to include both ancient and medieval influences

---

7   For example Bacon and Descartes saw themselves as calling for not only the displacement of Aristotelian Science but also the structures of methodology and logic associated with it.
8   This change was largely brought about by the Romantic Revolution, culminating with Hegel's contrast between Greek and Oriental traditions of thought. For a critique of this Hegelian reorientation, see Tibebu (2011).

on modern science.[9] However, the influences acknowledged were still largely European – they constituted what may be described as a Eurocentric tunnel connecting different historical ages of Europe, but largely insulated from trans-European influences.

Needham essentially broke the Eurocentric tunnel vision by showing that a second tunnel of influence fed into modern science – one that stretched from China in the east to Europe in the west through the corridor of communication created by the Mongol Empires. As the Needham series on Chinese science expanded it inspired similar efforts to record the contributions of other civilizations to modern science. These include S.H. Nasr's *Science and Civilization in Islam* (1968), and Bose, Sen and Subbayarappa's *A Concise History of Science in India* (1971). These dialogical projects have expanded, deepened and specialized over the last five decades so that there is now a large body of literature on the scientific contributions of various Asian cultures and civilizations to modern science.

However, earlier dialogical studies followed Needham's approach in taking a binary perspective where a single non-Western civilization is studied to see how it contributed to modern science in Europe. But more recent dialogical works have also begun to focus on understanding how philosophical, scientific and technological discoveries from various non-Western civilizations came to be combined in Europe to advance modern science. Thus, George Gheverghese Joseph in his study *The Crest of the Peacock: The Non-European Roots of Mathematics* shows how the growth of modern mathematics came to be shaped in Europe by flows of ideas from India, China and the Islamic-Arabic world. Similarly John Hobson in *The Eastern Origins of Western Civilization* also traces in detail how resource portfolios (ideas, technologies, institutions) from across the Asian continent were drawn by modern Europeans to create science, and the social, economic and political institutions associated with it. All of these studies build on the binary studies that were done in the previous five decades. They suggest that what was important was not the Greek heritage alone, since this was inherited by the Arabs, but also the reservoirs of knowledge from Asian traditions of science which Europeans incorporated into modern science.[10]

---

9   Both the periodization and existence of a Dark Age, as well as the conflation of the Medieval period with the Dark Age are highly contested notions. See Duhem (1985).

10   See also Bala (2006) who argues that even the Scientific Revolution of the seventeenth century drew on the astronomical heritage of the Arabic Maragha School critiques of Ptolemy, the Indian Kerala School mathematical techniques, and Chinese infinite empty space cosmology.

Such a multi-polar dialogical understanding of the historical origins of modern science requires us to reconsider Cohen's view that Needham's Grand Question is not relevant to the case of Southeast Asia, since this region did not have any science to begin with. However, he also assumes that the Romans did not either, but yet it was those who saw themselves as the descendants of the Roman Empire who produced modern science. Moreover, if we take into account connected histories of science, they accomplished this by drawing on the resources of the Asian civilizations with the most advanced sciences. This raises the question, why did Europe engage in a dialogue of civilizations to produce modern science despite the fact that it began with a heritage derived from Rome that Cohen considers to be no more likely to produce modern science than the heritage of Southeast Asia?

5   Why Europe?

To answer the question it is fruitful to look at what Needham has to say in his final statement on these matters in the chapter "General Conclusions and Reflections" which he wrote for his last volume contribution to the *Science and Civilization in China* series. Concerning the epistemology of science he writes:

> In *Science and Civilization* we have viewed science in three levels, first what we call proto-science, such as is found in Babylonia; then medieval science, such as is found in China before the year 1700 and Europe before 1500 approximately. Thirdly we refer to modern or international science.... I have pictured modern science as being like an ocean into which the rivers of all the world's civilizations have poured their waters. (Needham 2004: 201)

But Needham is also careful to note that although modern science received contributions from pre-modern traditions the two should not be confused:

> I believe in making a sharp distinction between modern science and that which preceded it, namely ancient and medieval science. Modern science, to my mind, consists of two things, the mathematisation of hypothesis about Nature on the one hand, and on the other continuous and relentless experimentation. In my opinion there is no science outside modern international science. (Needham 2004: 202–203)

Consider what Needham termed "mathematisation of hypothesis". This is intimately tied to the notion Galileo defended by claiming that mathematics is the language used to write the Book of Nature. Newton also made this mathematical hypothesis assumption when he claimed that his inverse square law of gravitation guided all bodies in the universe with absolute exactness and precision. The same belief inspires modern physicists who take the mathematical laws of quantum physics to predict the behavior of bodies as tiny as strings and as vast as the expanding universe. But mathematization of hypotheses was never part of Greek science. Neither Platonists nor Aristotelians subscribed to such an exacting conception of the applicability of mathematics to the world. The Egyptian-Greek mathematician Euclid of Alexandria only worked with ideal mathematical forms, such as perfect triangles and perfect circles, and did not assume that the relations he demonstrated were exemplified by physical objects in the world. A few centuries later his fellow Egyptian-Greek astronomer Ptolemy of Alexandria, developed a mathematical model of the universe that was to inspire Greek, Arabic and early modern European science, in which he also assumed that his theory was only designed to save the phenomena and not represent them with realist accuracy. Even Archimedes of Syracuse, who is often seen as the archetypal embodiment of Greek mathematical rationalism, worked with idealized representations of levers that moved upon frictionless fulcra, and pulleys that had no weight.

I would like to argue that the modern notion of mathematization of hypotheses only came to be seen as a possibility after Ibn al-Haytham's ray theory of optics came to inspire scientific thinkers in medieval Europe. His theory assumed that light rays traveled in perfect straight lines when unimpeded, and that their reflections and refractions could be viewed as obeying geometrical laws in all their exactness. Thus al-Haytham's theory of optics suggested that phenomena in nature could be read as perfectly conforming to mathematical relations and laws. By integrating mathematics with the optics of sensory experience al-Haytham laid the basis for the mathematical rationalism of modern science.[11] It also led to a transformation in European aesthetic sensibilities by inspiring the perspectivism of Renaissance art, as Hans Belting (2011) has recently argued, where European artists sought to represent the world in painting exactly as it appeared to the visual sense.

We also know from Needham's studies that the initial impetus for European technological development in recent times came from the flood of Chinese mechanical technologies into Europe during the medieval period. However, within China these technologies can be seen as inspired by Chinese correlative cosmological views closely linked to the Chinese organic materialist conception of the universe. Organic materialism fostered a tinkering orientation

---

11   See Bala (2006), Chapter 8 "The Alhazen Optical Revolution." This has also been systematically traced in a recent study by Belting (2011).

toward mechanical systems – one designed to examine how changing the behavior of one part of a system would affect other parts. This is, of course, the trial and error process often used to advance technology for millennia in Chinese culture – what may be called the 'correlative experimentalism' of Chinese culture.[12]

Francis Bacon's experimental method which inspired modern science can also be seen as taking to a higher level Chinese correlative experimentalism. Unlike the Chinese, however, Bacon recommended going beyond the mere tinkering orientation by creating new contexts which did not exist in nature – torturing nature, as he put it – in order to discover those invariant correlations (what we now call laws) that ruled in all possible experimental contexts. Thus it is possible to see the inspiration provided by Chinese technologies, and the correlative experimentalism associated with them, as the source for Bacon's 'discovery of how to discover'.

Hence what Needham refers to as mathematization of hypothesis and relentless experimentation is grounded in influences rooted in the Arabic tradition of Alhazen optics and Chinese technological experimentalism. These came to be combined in the era that has been seen as the dawn of the modern world – the European Renaissance, when the influences from the two non-European civilizations came to meet in Europe.

In order to recognize how this happened we have to go beyond the standard view of the Renaissance which assumes that it involved the rebirth of Hellenic and Hellenistic philosophy, literature, arts and sciences in the soil of Western Europe after an interregnum of nearly a millennium. In that period, often seen as the era of the so-called Dark Ages, the ancient heritage is assumed to have been adopted and carried by the Arabic civilization which, although it held it in high regard, did not contribute any further significant advances to it. When this heritage was returned to its rightful heirs in Europe, it became energized once again and led to those intellectual advances that produced modern science and philosophy.

This argument ignores the fact that the region in which the Renaissance developed in Europe was for three centuries prior sandwiched between two major influences that emanated from the Iberian Peninsula in Western Europe and Russian territories in Eastern Europe. The Iberian territories were under Islamic rule for over seven centuries from the eighth to late fifteenth century when the last Muslim stronghold in Europe, Granada, fell to the *Reconquista* in 1492, in the same year that Christopher Columbus landed in America. In Russia, Mongol sovereignty, albeit erratic and fluctuating, began in the first half of the thirteenth century and ended only in 1480.

---

12   For a more systematic discussion of these issues see Bala (2008), pp. 126–130.

Hence during the two and a half centuries from the Scholastic Age to the Italian Renaissance medieval Europe came to be sandwiched and open to the influence of the two most scientifically advanced civilizations in the world at that time. Mongol rule in China and the Mongol control of Russia and the intervening territories to China, opened a corridor of communication from China to the West along which flowed a whole stream of Chinese technologies created over a millennium – a process that Needham has amply documented.

This influence from the East was complemented by another from the West as the *Reconquista* opened up access to the scientific works of leading Arabic thinkers such as Avicenna, Averroes, Alhazen and Rhazes whose impact on a receptive Europe cannot be underestimated. The Arabic influence was particularly significant in the contribution made by Alhazen optics to the development of a mathematical perspectivism and realism that informed Renaissance art.

In many ways, it is possible to argue, that it was the fusion of these two streams in Middle Europe that constituted the main scientific influence on the Renaissance. Even an iconic Renaissance figure such as Leonardo da Vinci can be seen as reflecting these dual influences in his interest in the application of perspective mathematics in art, and his interest in mechanical engineering – they reflect the dual Arabic and Chinese influences on the European Renaissance.

In his introduction to the final volume of the Needham series on *Science and Civilization in China* Mark Elvin indirectly points to the power of combining Arabic mathematical realism with Chinese technological experimentalism when he writes:

> The Chinese artisans who built rectangular trough pallet-chain pumps for drainage and irrigation in the lower Yangtze region in late imperial times altered the proportions of troughs and pallets by (one assumes) trial-and-error to optimize the use of energy for different angles of inclination of the trough, presumably knowing but keeping to themselves the appropriate empirical proportions. The 18th century French hydraulic engineer De Belidor, using simple Euclidean geometry plus simple mechanics, calculated and published the specific optimal ratios for the same type of pump – which was clearly, by a nice twist, borrowed from China. (Elvin 2004: xxxii)

Here then we have a clear-cut answer to the question why Europe, rather than Southeast Asia, produced modern science. Unlike Europe which drew on the scientific achievements of Arabic and Chinese civilizations, Southeast Asia did not become an important zone for the dissemination of either Arabic scientific ideas and theories or Chinese technologies. Moreover, the optics of Alhazen did not become introduced into China, nor did Chinese technologies have the same impact on Islamic civilization. Thus in none of the other civilizations did the mathematization of hypotheses come to be combined with technological experimentalism. We have to conclude that the values of mathematization of hypotheses and technological experimentalism that Needham considers to be crucial for modern science are really the result of Europe's dialogical interactions with the scientific heritage of the two most advanced scientific civilizations of the Muslims and Chinese with which it came to be connected along its Western and Eastern borders during the Bright Dark Ages.

We began this paper by examining the views of Cohen and Huff that assign a central role to Aristotelian philosophy as paving the way for modern science to emerge in Europe. However, it is also noteworthy that modern science developed largely in opposition to Aristotelian conceptions by drawing on mechanical technologies and the mechanical philosophy they inspired. Moreover, the consolidation of mechanical science came to be possible by drawing on technological and theoretical discoveries outside the West – especially the Chinese and Islamic civilizations. This suggests that any answer to the Needham question must go beyond the comparative studies of science and technology in China, Islam and the West adopted by Cohen and Huff. In order to account for the rise of modern science in Europe we must also take into account the flows of knowledge across these civilizations that connected them. It is precisely for this reason that comparison between Southeast Asia and Europe is as relevant for answering the Needham Question as comparisons with China, India and the Islamic world.

Bibliography

Al-Andalusi, Said (1029–1070). *Science in the Medieval World: Book of the Categories of Nations*. Translated by Sema'an I. Salem and Alok Kumar (1991). Austin: University of Texas Press.

Bala, Arun (2006). *The Dialogue of Civilizations in the Birth of Modern Science*. New York: Palgrave Macmillan.

Belting, Hans (2011). *Florence and Baghdad: Renaissance Art and Arab Science*. Belknap Press of Harvard University Press.

Boruchoff, David A. (2012). "The Three Greatest Inventions of Modern Times: An Idea and Its Public" in Klaus Hock and Gesa Mackenthun (eds.) *Entangled Knowledge: Scientist Discourses and Cultural Difference*. Munster and New York: Waxmann, pp. 133–163.

Bose, D.M., S.N. Sen, and B.V. Subbarayappa (1971). *A Concise History of Science in India*. New Delhi: Indian National Science Academy.

Cohen, H. Floris (1994). *The Scientific Revolution: A Historiographical Inquiry*. Chicago: University of Chicago Press.

Cohen, Floris (2011). *How Modern Science Came into the World: Four Civilizations, One 17th-Century Breakthrough*. Amsterdam University Press.

Duhem, Pierre (1985). *Medieval Cosmology: Theories of Infinity, Place, Time, Void and the Plurality of World*. Translated by R. Ariew. Chicago: University of Chicago Press.

Elvin, Mark (2004). "Vale Atque Ave", in Needham (2004), pp. xxxiv–xliii.

Hobson, John M. (2004). *The Eastern Origins of Western Civilisation*. Cambridge University Press.

Huff, Toby E. (2003). *The Rise of Early Modern Science: Islam, China, and the West*. Cambridge: Cambridge University Press.

Joseph, George Gheverghese (2000). *The Crest of the Peacock: Non-European Roots of Mathematics*. Princeton & Oxford: Princeton University Press. First published by I.B. Tauris (1991).

Lieberman, Victor (2003). *Strange Parallels: Southeast Asia in Global Context, c.800–1830*. Yale University Press.

Lindberg, David (1992). *The Beginnings of Western Science: The European Scientific Tradition in Philosophical, Religions, and Institutional Context, 600 BCE to AD 1450*. Chicago: University of Chicago Press.

Low, Morris F. (ed.) (1999). *Beyond Joseph Needham: Science, Technology, and Medicine in East and Southeast Asia*. Chicago: Chicago University Press.

Nasr, Seyyed Hossein (1968). *Science and Civilization in Islam*. Cambridge, Mass.: Harvard University Press.

Needham, Joseph (2004). *Science and Civilization in China*, vol. 7(ii), *General Conclusions and Reflections*. Cambridge: Cambridge University Press.

——— (1970). *Clerks and Craftsmen in China and the West*. Cambridge: Cambridge University Press.

——— (1969). *The Grand Titration: Science and Society East and West*. London: Allen & Unwin.

Pacy, Arnold (1990). *Technology in World Civilization: A Thousand Year History*. Cambridge, Mass.: MIT Press.

Raj, Kapil (2007). *Relocating Modern Science: Circulation and the Construction of Scientific Knowledge in Southasia and Europe*. New York: Palgrave Macmillan.

Reid, Anthony (1990). *Southeast Asia in the Age of Commerce, 1450–1680*. Yale University Press.

Ronan, Colin (1983). *The Cambridge Illustrated History of the World's Science*. Cambridge: Cambridge University Press.

Saliba, George (1994). *A History of Arabic Astronomy: Planetary Theories during the Golden Age of Islam.* New York and London: New York University Press.

———— (2007). *Islamic Science and the Making of the European Renaissance.* Cambridge, Mass.: The MIT Press.

Tibebu, Teshale (2011). *Hegel and the Third World: The Making of Eurocentrism in World History.* New York: Syracuse University Press.

Yu, Jiyuan (2007). *The Ethics of Confucius and Aristotle.* New York: Routledge.

CHAPTER 12

Rethinking the Needham Question: Why Should Islamic Civilization Give Rise to the Scientific Revolution?

Mohd. Hazim Shah[1]

1   Introduction

The Needham question as applied to Islamic Civilization appeared in the form "Why Didn't the Scientific Revolution Occur in Islamic Civilization"? The question has been discussed by, among others, Pervez Hoodbhoy (1992: 118f.), Toby Huff (1995: 47f.), Floris Cohen (1994: 381–83), and S.H. Nasr (1996: 130f.). The question of course, including the original Needham question which was applied to Chinese Civilization, contains a presupposition; that the Scientific Revolution is a good thing. In this paper however, I take a different perspective on the Scientific Revolution, and instead choose to look at the cultural problematic brought about by the Scientific Revolution rather than its touted success. In this regard I refer to recent critical historiographies of the Scientific Revolution, for example by Andrew Cunningham (1993), Stephen Gaukroger (2006), and Peter Dear (2005). Such critical historiographies are perhaps not unrelated to the critical reaction against science found in writings on the history and philosophy of science of the last two decades, including those linked to the so-called 'science wars'. Unlike the previous historiography of the Scientific Revolution, associated with Herbert Butterfield (1957[1949]) and Rupert Hall (1954), the new historiography does not privilege science. So in the context of the new historiography the force of the original question gets somewhat dampened, if not made irrelevant. However, I will use the new historiography not as a means of debunking the question altogether, but rather to enable us to look at the 'other side' of the Scientific Revolution, which was previously not possible because of the prestige and privileged position accorded to science in the old discourse. But now with there being efforts made to

1 An earlier version of this paper was presented at the Conference on "The Bright Dark Ages", organized by Institute of Southeast Asian Studies (ISEAS), Singapore, 20–21st May 2010. I would like to acknowledge the University of Malaya for providing me with the UMRG research grant number RG437-13HNE to carry out this research.

re-examine science, and the quest for alternatives to modern science, it is not inappropriate for us to start with a diagnosis of the source of the cultural problematic brought about by modern science, and I choose to locate that source in the rise of modern science, i.e. in the Scientific Revolution itself.

I will look at recent theories of the Scientific Revolution which attributed the rise of modern science to the combination of three major components, namely natural philosophy, mathematics and the experimental method.[2] I will then discuss the significance of the fusion of the three elements in the rise of modern science, and relate it to the question of 'disenchantment'. The fusion of theory/world-view and practice implicit in the transformation, together with its significance and cultural implications will also be discussed. I will then look at how both the West, and later Islam, responded to the cultural problematic which occurred as a result of the rise of modern science. I will argue that although Islam did not give rise to the Scientific Revolution, the scientific presence in Islam before the modern era, unlike in post 17th century Europe, did not create the cultural and intellectual conflict which afflicted Europe and which it is still grappling with. Here I will draw on the works of the historian of Islamic science, Seyyed Hossein Nasr (1976, 1988, 1996). By looking in a comparative manner, at the different trajectories which science took in Islam and the West, it is hoped that some light can be shed on the contemporary cultural problematic of science and its relation to modernity which affects us all today.

2   The Needham Question and the New Historiography of the Scientific Revolution

The Needham question in the form, "Why Didn't the Scientific Revolution Occur in Civilization X?" was posed amidst the old historiography of the Scientific Revolution, associated with A.R. Hall (1954) who wrote the book entitled *Scientific Revolution*, and Herbert Butterfield (1957[1949]) who wrote *The Origins of Modern Science*, and was later given credence through the notion of 'scientific revolution' as it was expounded in Kuhn's (1970[1962]) classic, *The Structure of Scientific Revolutions*. Since then, the historiography of the Scientific Revolution has been subjected to critical scrutiny and revision, so much so that the very notion of 'the Scientific Revolution' as it was

---

2   See Cohen (2005, 2010), Gaukroger (2006), Grant (2007).

propounded by Butterfield and Hall, has lost its earlier appeal and grip on the historian's mind.[3]

What are the issues concerning the Scientific Revolution raised by the new historiography? Some of these are:

1. The issue of whether seventeenth century developments in science can really be considered 'revolutionary'? This is posed in terms of the 'continuity thesis' versus the 'revolutionary thesis' – whether the theoretical and methodological changes from medieval to modern science are evolutionary or revolutionary (Lindberg 1992). The continuity thesis had in fact been propounded earlier by Duhem through his study of medieval science, especially the science of mechanics. More recently, it was advocated by A.C. Crombie who argued for a methodological continuity between medieval and modern science. Supporters of the concept of 'scientific revolution' such as A.R. Hall, I.B. Cohen, and Thomas Kuhn, however insist on the revolutionary nature of the changes in 17th century science.

2. A second issue, forcefully argued by Andrew Cunningham and Perry Williams (1993) in their influential article entitled 'De-centering the Big Picture: The Origins of Modern Science and the Modern Origins of Science', insisted that the very notion of a 'scientific revolution' implies the acceptance of the central role of science in the shaping of modernity, and a surreptitious way of privileging science in modern culture. This is also related to the issue of 'Whiggism' in the historiography of science. It also fits in with recent postmodernist critiques of science, in which science is not seen as epistemologically or cognitively privileged *vis-à-vis* other knowledge or belief-systems. According to this view, historians should no longer entertain the notion of 'the Scientific Revolution' because of its philosophically unacceptable assumptions and underpinnings.

---

3 The concept of the 'Scientific Revolution' first found in the writing of Butterfield (1949) and then Hall (1954), later went through a period of shifting fortunes. It came for critical appraisal and review in the book edited by Lindberg and Westman (1990) entitled, *Reappraisals of the Scientific Revolution*, and was criticized by Cunningham and Williams (1993), and Dear (2005). Such a critical view of the Scientific Revolution as modernity's 'myth of origin' can also be found in Park (2006: 15). The concept also has its defenders, namely Westfall (1978), Biagioli (1998), and most recently Cohen (2010), whose very work depended on the viability of the concept. However, in my assessment, the concept no longer privileges science in the way it used to before.

3.   The third issue concerning the historiography of the Scientific Revolution
     relates to the very notion of "science" itself, as found in the writings of
     Joseph Rouse (1993) and Peter Dear (2005). Both Rouse and Dear argue
     that the notion of a unitary entity called "science", persisting in a stable
     manner through time, is open to question. So if 'science' as such a unitary
     entity does not meaningfully exist, then talk of 'the Scientific Revolution'
     is equally meaningless.

Except for (1), (2) and (3) are really post-modernist philosophies and historiography critical of science and modernity, in which the notion of the Scientific
Revolution is seen as supporting a modernist image of science and its legitima-
tion. Now, given this new critical historiography of the Scientific Revolution,
what are we to make of the Needham question, which partly derived its force
and significance from the old historiography? Doesn't it seem blatantly mis-
placed? Or should we reject the new historiography – calling it 'postmodernist
prejudice' – and insist on the relevance of the Needham question? I do not
think that we can lightly dismiss the new historiography, for it does bring out
in a sense the prejudices underlying the old historiography, which prevented
a fair assessment of the cultural role of science in society. For the old historiography immediately leads us to apportion blame on those civilizations that
did not produce modern science, as can be seen in the various attempts at
teasing out the factors that existed then in those civilizations, which were held
to be responsible for the non-emergence of science.[4] There was hardly any
attempt at asking whether there were problems associated with the Scientific
Revolution and the rise of modern science, unlike the critical questions that
were raised concerning the Industrial Revolution of the 18th century. Now it
is to the credit of the new historiography, that by 'de-centering the Big Picture
of the Scientific Revolution',[5] we can now adopt a more open attitude towards

---

4   See for example, Huff (1995) and Cohen (1994).

5   Cunningham and Williams (1993: 430) express the idea of 'de-centering' as non-privileging as
    follows: "In a more general sense, de-centering is something which we continue to do repeat-
    edly throughout our adult lives, as we identify yet another aspect of our own egotism, and
    realize that something which we thought was universal is actually peculiar to ourselves, or
    our group, our class, our nation, or our culture. To see science as a contingent and recently-
    invented activity is to make such a de-centering and to acknowledge that things about our
    primary way-of-knowing which we once thought were universal are actually specific to our
    modern capitalist, industrial world". Thus the idea promoted by the concept of the 'scientific
    revolution' that science is a unique product of western civilization, that is universally true, is
    rejected with the realization that it is the result of a unique set of historical circumstances –

the development of science in Western as compared to non-Western cultures and civilizations.

3   On why Islamic/Arabic Science Did Not Give Rise to the Scientific Revolution: The Views of Pervez Hoodbhoy, Toby Huff, and A.I. Sabra

The classical version of the Needham question had been posed for science in Islamic civilization by several scholars, namely Pervez Hoodbhoy (1992), Toby Huff (1995), A.I. Sabra (1987), and H. Floris Cohen (1994), who each have their own view as to why Islamic civilization did not give rise to the Scientific Revolution, despite having advanced the sciences which they inherited from the Greeks. Hoodbhoy, while acknowledging the significant contributions made by the Muslims in science and technology during the Middle Ages, nevertheless laments the decline of science in the late Middle Ages. Hoodbhoy (1992: 118–133) mainly attributes the decline of science in Islam to values and institutional factors, such as its legal and education system, which to him, does not encourage creativity and rational thought. He pointed out 'rote learning' in Islamic education as an example.

Toby Huff (1995) in his comparative study of science in Islam, China and the West, also tried to identify the reasons for the non-emergence of modern science in Islamic civilization despite it having assumed scientific leadership from around the 9th to 12th centuries. Huff looks at the question from a sociological point of view, and in terms of modernization theory. He draws upon the sociological theories of Max Weber and Benjamin Nelson with regards to modernity. For Huff, Islamic society lacks the value-system and institutional structures that would allow for the further progress of science and the rise of a scientific revolution (Huff 1995: 212–13). He harps on the question of "rationality" (Huff 2000), and of the 'institutional impediments' to the rise of modern science. Huff has been criticized for his 'Whiggish' approach to the history of science, and for pre-judging the question of modernity in the history of science (Lloyd 1997). Huff's use of the Mertonian norms (Huff 1995: 213) in his diagnosis of medieval Islamic science[6] is not only an anachronism,

the product of the Age of Revolutions in general – peculiar to European history. Thus its universal appeal is explained not in epistemological, but in historical and sociological terms.

6   Huff (1995: 213) wrote: "If the scientific worldview is to prevail, its elements of universalism, communalism, organized skepticism, and disinterestedness must be given paradigmatic expression in the dominant directive structures of a society. A major clue as to why Arabic

but also impervious to the criticisms that have been leveled against Merton's norms. Like Hoodbhoy, he did not dismiss the important contributions made to the development of science by scientists in the medieval Islamic world. So while both Hoodbhoy and Huff acknowledged the cognitive capacity of Muslim scientists to advance the sciences – which explains their significant contributions – they nevertheless believe that there are factors in medieval Islamic culture and civilization such as values and institutions, that did not allow for the full flourishing of science in Islam. However, because they have already pre-judged the question of modernity and the positive value attached to it, their asking of the Needham question inevitably leads them to put blame on Islamic culture for not possessing the characteristics that would allow for the emergence of modern science. Thus their query is not neutral and largely shaped by the old historiography and the privileging of modernity.

Sabra on the other hand, adopts a more charitable attitude towards Islamic science in his 1987 paper published in the journal *History of Science*, which bears the title 'The Appropriation and Subsequent Naturalization of Greek Science in Medieval Islam: A Preliminary Statement'. As opposed to the 'marginalization thesis' propounded by Von Grunebaum (see Lindberg 1992; Cohen 1994: 389–394), which claimed that the position of science (or *'ilm awwal*) was marginal in Islamic culture and education, and that it was not part of mainstream Islamic thought or education, Sabra argued that science had a more secure and entrenched position in Islamic society. His thesis was dubbed the 'appropriation thesis' because it states that science was appropriated into Islamic culture and society, especially in the case of mathematics and astronomy, which served religious functions such as determining the direction and time of prayer. Sabra (1987: 240–42) however, lamented on the 'instrumentalist' as opposed to the 'realist' character of Islamic science, and argued that the significant achievements made by Islamic science was because of its instrumentalism.

However, instrumentalism understood as 'a special view which confines scientific research to very narrow and essentially unprogressive ideas' (Sabra 1987: 241) could only bring about limited progress and not lead to scientific breakthroughs.

---

science failed to give birth to modern science can therefore be found in the fact that these norms were not institutionalized in the directive structures of Islamic civilization". The Mertonian norms have been challenged and debated, and is seen by some as a relic of the Cold War. One common argument against it is the rise of science in socialist countries of the 1960s and 1970s, where the norms were not institutionalized. Thus Merton's norms could not have been a necessary condition for the rise of modern science.

The reason which Sabra offered for this 'limited progress' is because the *hakims* or the 'philosopher-scientists' are not allowed free rein in an Islamic culture, since this would put them on a collision course with the religious orthodoxy. According to Sabra (1987: 240):

> The philosophical sciences ... had entered Islamic intellectual life under the banner of an articulated concept of *hikma* (or philosophical wisdom) which involved a doctrine of knowledge quite distinct from [an instrumentalist and religiously oriented view]. According to this doctrine, the aim of theoretical investigation, whether mathematical, physical or metaphysical, was to ascertain the nature of all things as they are in themselves ... The metaphysically inclined seeker after the truth, namely all the great philosophers of Islam and the majority of mathematicians in the earlier period, also believed that the ultimate value of all science was to perfect the human soul and prepare it for a state of eternal happiness – that is the very same state which Ghazali hoped to secure through religious knowledge. But for the philosopher and the philosophically committed scientist, this was not a state that transcended philosophical activity ... Man's perfection thus lay in the perfection of his philosophical or scientific knowledge and the way to his salvation was none other than the way of science. By contrast, a logical consequence of Ghazali's view was, as we have seen, to put a curb on theoretical enquiry.

In other words, the Muslim philosopher-scientist could not openly adopt a realist attitude towards science. Instrumentalism as a philosophical attitude, does not contribute towards the development of science in Islamic Civilization as such, but it does help to steer away from murky epistemological problems which would arise if science were to be adopted in a realist sense, because then it would stand in competition with religion as the purveyor of true knowledge about the world. Thus keeping an instrumentalist attitude towards science enabled the full potential of science to be developed within that framework. But to go beyond it, as Grant (2004, 2007) rightly notes, would require the inclusion of natural philosophy in order to broaden the theoretical base. But since natural philosophy is closely related to realism, this did not become a serious option.

In so arguing, eventually Sabra does not differ essentially from Hoodbhoy and Huff, who both argue for the stronger 'rationalist-realist' thesis in the rise and advancement of modern science. It now appears that the heart of the issue concerning the non-emergence of science in Islamic civilization revolves on the question of rationality and realism, and its institutional expressions.

4   The Integration of Natural Philosophy, Mathematics and the
       Experimental Method and the Rise of Modern Science in Europe
       in the 17th Century

Recent studies on the Scientific Revolution and the rise of modern science have
tended to focus on the integration of natural philosophy, mathematics and the
experimental method (Henry 2008; Cohen 2005, 2010; Gaukroger 2006; Dear
2005, 2006; Grant 2007). Natural philosophy existed in separation from mathematics, the exact/mathematical sciences, and the experimental method prior
to the emergence of modern science in the 17th century. It is to this crucial
historical phenomenon in the rise of modern science, which this section will
be devoted to. Apart from the various theories that have been put forward in
the past on the causes of the Scientific Revolution (Hall 1983), the new account
which concentrates on the fusion or amalgamation of natural philosophy,
mathematics and the experimental method serves to further our understanding of the rise of modern science, regardless of the historiographical issues
surrounding the 'Scientific Revolution'. In the old historiography, these theories were classified as either 'internalist' or 'externalist'. Examples of externalist accounts are the ones put forward by Robert K. Merton, Edgar Zilsel, Boris
Hessen, and Lewis Feuer, while the internalist thesis is associated with historians such as A.R. Hall and Alexander Koyre (see Basalla 1968). This new account,
however, cannot be classified as being either strictly 'externalist' or 'internalist',
since it includes elements of both in an interesting way. Although mathematics, and the experimental method, are larely 'internal' to science – notwithstanding the contrary claims made by Steven Shapin and Simon Schaffer (1985)
with regards to experimentation – are largely 'internal' to science, the status of
'natural philosophy' is somewhat ambiguous. To claim that natural philosophy
is 'internal' to science, would be somewhat Eurocentric, since its acceptance
is not universal, as can be seen in the case of Islamic science. If anything, it
indicates that the 'cultural' can be constitutive of the construction of science
itself, rather than being merely 'contextual'.
      The new accounts of the Scientific Revolution offered by the above-
mentioned scholars are by no means monolithic and in complete agreement
with one another. Edward Grant (2007) for example, wrote his book *A History
of Natural Philosophy*, partly to counter the views put forward by Andrew
Cunningham and Perry Williams (1993) – dubbed as the 'Cunningham thesis'
by Peter Dear (2001) – on the relationship between natural philosophy, theology, and natural science. For Cunningham, the emergence of natural science
involved the abandonment of natural philosophy (which to him occurred as

late as the 18th or 19th century), and that the two did not co-exist. Furthermore to him, natural philosophy in Europe was all about God's creation (Grant 2007: 250). According to Cunningham (1993: 421):

> For the whole point of natural philosophy was to look at nature and the whole as created by God...Natural philosophy scrutinized, described, and held up to admiration the universe as the true God had created it and kept it running. To the modern ear, accustomed to the distinction between 'science' and 'religion', and to a clear-cut distinction between the 'sacred' and the 'secular', this may sound as though natural philosophy was merely an aspect of theology (and particularly that it was 'natural theology'). But this was not the case: natural philosophy was an autonomous study separate from theology and from natural theology.

Thus for Cunningham it was clearly the case that natural philosophy – although distinct from theology – was about the study of nature as God's creation, and hence his judgment that it ceased to exist after the seventeenth century.

Grant on the other hand believed that natural philosophy and the 'mixed sciences' (which resembled our contemporary 'natural science') did co-exist in the past, and that they became amalgamated after the seventeenth century. This difference of opinion between him and Cunningham is highlighted in the following passage (Grant 2007: 303):

> Andrew Cunningham and Perry Williams have claimed that natural philosophy and science were two wholly different enterprises that never coexisted: natural philosophy appeared first in the ancient world and continued to exist for many centuries until it was wholly replaced by science, or modern science, in the nineteenth century...I shall argue...that these are serious misunderstandings of the history of science. The Scientific Revolution occurred because, after coexisting independently for many centuries, the exact sciences of optics, mechanics, and especially astronomy merged with natural philosophy in the seventeenth century. This momentous occurrence broadened the previously all-too narrow scope of the ancient and medieval exact sciences, which now, by virtue of natural philosophy, would seek physical causes for all sorts of natural phenomena, rather than being confined to mere calculation and quantification. Thus were the seeds planted for the flowering of the modern version of the exact physical sciences, and the many other sciences that emerged during the eighteenth and nineteenth centuries.

Whatever the differences in view and position concerning the relationship between natural philosophy, mathematics and the experimental method, they are all in agreement on one point; namely that an understanding of the rise of modern science must take into account the changing relationship between the three domains from the Middle Ages to the 17th century. On the finer question of whether natural philosophy still survived after the 17th century, or in what form, that seems to leave room for debate. For example an argument can be made that natural philosophy survived in the form of the 'mechanical philosophy' in the 17th century and after, through the Cartesian world-picture.

One interesting group of sciences that has existed in the past, and that served as a 'precursor' to modern natural science, is the 'mixed sciences' which includes disciplines such as astronomy, statics, mechanics, and optics. They are called 'mixed' because they contained elements of both mathematics and explanatory first principles (which might include natural philosophy). Not surprisingly, the 'mixed sciences' have also been the subject of enquiry by historians such as Kheirandish (2006), Grant (2007), Gaukroger (2006), and Cohen (2010), who are interested in the phenomenon of the emergence of modern science. Questions are here posed such as: (i) what is the relationship between the mixed sciences and natural philosophy? (ii) where do the mixed sciences stand in relation to mathematics and natural philosophy? (iii) how do the 'mixed sciences' differ from modern natural science? (iv) do the mixed sciences such as astronomy, give rise to issues such as 'scientific realism'? That the mixed sciences – related more to the Alexandrian tradition which was operational – was pursued in Islamic civilization serves to further raise food for thought on why no 'scientific revolution' occurred in Islam.

The relationship between religion, natural philosophy, the exact sciences and the experimental method took different paths in Islamic and Western civilizations. In the West, there was an integration of natural philosophy, mathematics/exact sciences, and the experimental method which led to the rise of modern science in Europe in the seventeenth century.[7] This integration was partly facilitated by the religious accommodation of science in the West

---

7 The emergence of modern science in the seventeenth century, though influenced by its social and cultural context, was still largely independent of technology and the economy. It was not until the nineteenth century that we begin to see the connection between science, technology and the economy through the Second Industrial Revolution. The relationship deepened in the twentieth century, which led to the notion of "Technoscience", and claims about the 'primacy of practice' in the analysis of science. However, as for the 17th century and before, we can safely assume that technology and the economy did not significantly figure in the evolution of scientific knowledge. See Forman (2007).

despite cultural strains as epitomized by the Galileo episode. Even as late as the early eighteenth century, we still find the great physicist Isaac Newton speaking of God in relation to the natural world. In Islamic civilization however, due to the lack of development of natural philosophy, such integration did not take place. Islamic science maintained the division of what Floris Cohen (2010) called 'Athenian' (natural philosophy) and 'Alexandrian' science and did not fuse the two. It developed mainly in the Alexandrian mode which is basically instrumentalist in nature. Such an approach enabled science to be pursued in Islamic Civilization, without serious religious objection.

5   Cultural Implications of the Rise of Modern Science

Despite the generally positive evaluation given to the rise of modern science in the West, it did not come without problems. In this section I will discuss the cultural problematic of the rise of modern science. My diagnosis of the problem really hinges on three things. One, the transformation of natural philosophy and the consequent transformation from an 'organic' to a 'mechanical' view of nature. Two, the integration of natural philosophy with mathematics and the experimental method, and the further loss of 'enchantment'. Three, the implications of the rise of modern science on the religious view of the world. My contention is that the rise of modern science created a profound shift in the cultural and religious outlook of the West which created problems that it is still grappling with today.

Let me start with the second point, i.e. the integration of natural philosophy with mathematics and the experimental method which essentially gave birth to modern science. Now, with Islamic civilization, due to the stunted development of (Greek) natural philosophy (Grant 2004: 237–243), its transformation into modern science as occurred in the West, did not happen. In the West, the transformation of natural philosophy and its fusion with mathematics and the experimental method finally saw the birth of modern science. Perhaps Francis Bacon was the philosopher who successfully captured the spirit and vision of this new science in his writings. However, it did not come without a price. The price paid was in terms of the 'loss of symbolism' including religious symbolism, which is essential in giving meaning and cultural value to human life. Or to be more precise, it is the loss of one's preferred cultural symbolism at the expense of that sanctioned by empirical science. Before, when natural philosophy was separated from practice, one had a choice of which natural philosophy to adopt. This gives the freedom to align one's culturally-oriented view of the world with technological practice. As Nasr

(1996: 126) puts it, previously 'traditional sciences of the cosmos [i.e. natural philosophy]...shared, in contrast to modern science, the same universe of discourse with the religion or religions of the civilization in whose bosom they were cultivated'. With the rise of modern science, this relationship has been torn asunder.

The cultural problem brought about by the emergence of modern science has sometimes been described in terms of the idea of 'disenchantment' associated with the German sociologist Max Weber (Gerth and Mills 1958: 139). Modern science is said to have brought about a 'disenchanted' view of the world in which nature is stripped of the Divine and the symbolic, and instead treated as a purely physical object to be analyzed in terms of space and time. However such a dis-enchanted vision of the world could not have emerged in Islam. This is because while Greek natural philosophy was successfully accommodated into Western thought and culture through a long process which spans several centuries, and was institutionalized in Western centers of learning such as the medieval universities, this was not the case in Islam. This does not mean however, that there was an *absence* of Greek natural philosophy in Islamic thought. Grant (2007) for example, has provided evidence for the existence of Greek natural philosophy in the works of Muslim philosophers such as Al-Kindi, Al-Farabi, Ibn Rushd, Al-Razi, and Ibn Sina. However, it did not enter into the mainstream of Islamic thought, was not institutionalized in Islamic centers of learning, and was opposed by powerful and influential theologians such as al-Ghazali. There was also no equivalent of a class of 'theologian-natural philosophers' as found in Europe (Grant 1996: 174–75). Greek sciences, termed as *'ilm awwal* or ancient learning, was marginal to Islamic thought and education. The kind of Thomistic synthesis carried out by St. Thomas Aquinas who synthesized Aristotelian philosophy with Christian theology in the West, was not only absent, but also unimaginable in Islam. So even if there was no political disintegration in Islam, as happened with the downfall of the Abbasid empire/caliphate, and other external factors had remained favorable, the Scientific Revolution would not have occurred in Islam because of the lack of development of natural philosophy.

The kind of science cultivated in Islam, and for which the Muslims were famous for, were the ones which were not connected with natural philosophy and were more applied in nature. Thus mathematics, medicine, astronomy, and optics were well developed in Islam. The 'mixed sciences' such as astronomy, optics and the mathematical sciences, which were more practical in nature were accepted and developed in Islam. Epistemologically, the attitude of the Muslims towards the sciences was instrumentalist rather than realist in nature (Sabra 1987). Realism and natural philosophy (in its Greek form) remained alien to Islamic spirit and culture. This precluded Islamic culture from adopting the path of a rationalized world of nature which only a realist natural philosophy could provide. Realism, of course, would have made theorizing more purposeful and committed since one is bent on finding 'the truth about the natural world', as in the case of Kepler. However, the decision to steer away from a realist natural philosophy saved Islamic culture from the kind of cultural tension due to the fusion of natural philosophy and instrumentality (Dear 2005). Such a fusion led to a realist conception of science, since its instrumentality is taken as a sign of the "truth" of its ontology. By rejecting Greek natural philosophy, and hence of the possibility of an alien ontology and metaphysics replacing the Islamic worldview, Islamic Civilization was able to retain its own metaphysical and enchanted view of the world. But this too came with a price: Islam did not acquire the power of materiality potentially present in the mapping of Logos (read 'theoretical resources of natural philosophy') onto nature, i.e. rationalization. However, it saved its soul from a cultural conflict which later gripped the West. This conflict was initially symbolized and epitomized by Galileo's struggle against the Roman Catholic Church, where Galileo took a realist attitude towards scientific theories, thus limiting religion's cognitive claim to truths about the natural world. More recently, this cultural crisis is manifested through debates on scientific realism.

As we saw in the case of the West, the scientific revolution was brought about by the fusion of natural philosophy, mathematics, and the experimental method. Although mathematics, the mixed or exact sciences, and even experimentation were found in Islamic civilization in the medieval period, natural philosophy was significantly absent. Without this crucial ingredient, even if the rest had continued on its ongoing trajectory, it is inconceivable that a scientific revolution would have been brought about.

6 Western and Islamic Responses to the 'Disenchantment' Brought About by Modern Science

Part of the literature in the history and philosophy of science can be seen as responses to the disenchantment brought about by modern science, and the attempt to reclaim enchantment through the humanities and social sciences (Schneider 1993). Joseph Rouse (1993) has located discourses in the philosophy of science as part of the discourse on modernity, in which contestations between modernists and postmodernists occur, and attempts to achieve re-enchantment is identified mainly with postmodernists. However, there are also philosophers of science such as Larry Laudan and Bas van Fraassen, who though not rejecting science and modernity, nevertheless reject a realist interpretation of science, and in so doing can be said to be resisting disenchantment. Thus it is possible to view the discourse and debates in history and philosophy of science concerning issues such as scientific rationality and scientific realism, as indicating the existence of a cultural problematic brought about by the rise of modern science, and its attempted resolution through historical and philosophical discourse.[8]

The Western response takes two major forms; the first is receptive of science and modernity while the second is not. Historians and philosophers of science who are receptive to science and modernity are generally both realists and rationalists. The second group, i.e. those who do not privilege science in modern culture, consists of the postmodernists and pragmatists or antirealists. The attempt to resist disenchantment, and to seek re-enchantment through the humanities, can be most clearly seen in the efforts of the second group. Their way of dealing with the cultural hegemony of science is to deny both epistemological and ontological privilege to science. They seek to show that science is not epistemologically privileged through a critique of scientific rationality, and that it is not ontologically privileged through a critique of scientific realism. The first group, i.e. the modernists, see the Scientific Revolution and the Enlightenment in positive terms and endorse both scientific rationality and scientific realism. As Gellner (1964: 179) puts it, 'science is the mode of cognition of industrial society', hence ensuring a consonance between science and modernity. The second group however, resists the encroachment of science into the cultural sphere. Looking at it in terms of Daniel Bell's (1996: 10) tripartite division of society into the three spheres, viz. the techno-economic, the polity, and the cultural, it is safe to say that while the first group endorsed science's presence and dominance in all three spheres, the second group would deny science's pre-eminence in the third, i.e. cultural sphere which is the only sphere where any semblance of 'spirituality' within a secular world can be retained.

The Islamic response on the other hand, though varied like its Western counterpart, shares a common element in not rejecting the role and place of religion in modern society, and its ready acceptance of the relevance of Islam in the modern world. Muslim attempts at re-enchantment can be found in the works of contemporary thinkers such as Seyyed Hossein Nasr (1988, 1996) and Syed Muhammad Naguib al-Attas (1978). Their efforts at re-enchantment however, draw upon the resources provided by Islamic philosophy, especially in relation to Sufism. This is especially evident in the writings of Nasr,

---

8 For a more detailed discussion of the issue, see Mohd Hazim Shah (2011).

who sought to retain the presence of the Sacred with regard to the natural world, and interprets modern science as giving us knowledge about 'lower realities', while knowledge of the 'higher realities' belong to the domain of Islamic or religious metaphysics. As Nasr (1976: 140) puts it:

> Altogether the Muslims made important contributions to various branches of mechanics and dynamics, departing in many ways from Aristotelian physics and even developing such cardinal concepts as momentum. They, however, never quantified physics completely nor ignored the symbolic nature of the natural world. Even their quantitative studies moved within the orbit of a cosmos which remained hierarchic, with each level of existence symbolizing the states above. They were moreover able to achieve what they did from the point of view of the history of physics without bringing about the calamities caused directly or indirectly by Galileo and his followers because the science developed by these Muslim figures always remained bound within the hierarchy of knowledge. The greatest of Muslim physicists like al-Biruni, Ibn al-Haytham and Qutb al-Din al-Shirazi accepted willingly this hierarchy and never attempted to make a quantitative science of the Universe central or relegate to the periphery the qualitative science of things which is the most essential precisely because at its highest level it alone can deal with the essences of things.

Another Muslim critic of modern science, Syed Naguib al-Attas (2001) also rejects the realist construal of scientific theories, reserving ultimate truths about nature to Islamic metaphysics alone. In comparison with their Western counterparts Nasr and al-Attas can be regarded as critics of modernity, except that their views are couched in an Islamic dress. However, because of the different position of Islamic societies with regard to the power-knowledge nexus, in comparison to the West, the Islamic rejection of science is never total. It stops at the point where scientific knowledge is clearly identified with Western power – both economic and military. Hence the Islamic position is dualistic at best – rejection of the metaphysics and ontology of science, while embracing its pragmatic and utilitarian aspects. This however, is not unlike the Muslim response to science and technology in the medieval past, in which Islamic science flourished in areas such as mathematics, astronomy, mechanics, and optics, i.e. the so-called exact or 'mixed sciences', which are utilitarian in nature and which served as handmaiden to religion, especially mathematics and astronomy. In fact, even until today, scientific realism is not an issue debated among Muslim thinkers, although admittedly it was discussed in the context of medieval astronomy. But even then, in the context of medieval astronomy, Islamic astronomy was largely 'instrumental' in nature, and not linked to natural philosophy (Sabra 1994). In this regard, it is a fair speculation to say that the Copernican revolution could not have occurred in Islam,[9] since the instrumentalist orientation of Islamic science would have precluded it.

7   Conclusion

Therefore, it is reasonable to conclude that although the arrested development of Greek or Greco-Arabic natural philosophy is a crucial factor in the non-emergence of modern science and the Scientific Revolution in Islam, the consequent instrumentalist character of Islamic science (though not without any realist or 'natural-philosophical' element as found in Ibn Haytham[10] for instance), nevertheless made it possible for Islamic civilization to avoid the cultural problematic of 'disenchantment' which confronted Western culture with modern science.[11] This makes disputable the relevance and significance of the Needham question for Islamic civilization. Indeed it is possible to rethink the question by asking "Why Should Islamic Civilization Give Rise to the Scientific Revolution?"

Bibliography

Al-Attas, Syed Naguib. 1993[1978]. *Islam and Secularism*. Kuala Lumpur: International Institute of Islamic Thought and Civilization (ISTAC).
———— (2001). *Prolegomena to the Metaphysics of Islam: An Exposition of the Fundamental Elements of the Worldview of Islam*. Kuala Lumpur: International Institute of Islamic Thought (ISTAC).
Al-Daffa, Ali A. and John Stroyls (1984). *Studies in the Exact Sciences in Medieval Islam*. Dhahran, Saudi Arabia: University of Petroleum and Minerals and Chichester: John Wiley & Sons.

---

9   See Hisham Ghassib, "Why Did Arabic Science Fail To Achieve The Copernican Revolution? Unpublished manuscript.
10   See Sabra (2003: 85) on the need to combine both "physical" (i.e. natural philosophy) and "mathematical" modes of argument, according to Ibn Haytham, in the study of optics.
11   For an articulate expression of how Islamic science 'saved its soul' and avoided the disenchantment of nature, see Nasr (1968, 1976, 1988, 1996).

Al-Faruqi, Ismail (1982). *Islamization of Knowledge: General Principles and Workplan.* Washington DC: International Institute of Islamic Thought.

Al-Hassan, Ahmad Y. (1996). "Factors Behind the Decline of Islamic Science After the Sixteenth Century," in Sharifah Shifa al-Attas (ed.), *Islam and the Challenge of Modernity.* Kuala Lumpur: International Institute of Islamic Thought and Civilization (ISTAC).

Anstey, Peter R. and John A. Schuster (eds.) (2005). *The Science of Nature in the Seventeenth Century: Patterns of Change in Early Modern Natural Philosophy.* Dordrecht: Springer.

Basalla, G. (1968). *The Rise of Modern Science: External or Internal Factors?* Lexington, Mass.: Heath.

Bell, Daniel (1996). *The Cultural Contradictions of Capitalism.* New York: Basic Books.

Biagioli, Mario. "The Scientific Revolution is Undead," *Configurations* (1998) 6(2): 141–148.

Blair, Ann. "Natural Philosophy," in Park and Daston (eds.) (2006), pp. 365–406.

Butterfield, Herbert (1957[1949]). *The Origins of Modern Science.* London: G. Bell.

Cohen, H. Floris (1994). *The Scientific Revolution: A Historiographical Inquiry.* Chicago: University of Chicago Press.

——— (2005). "The Onset of the Scientific Revolution: Three Near-Simultaneous Transformations," in Peter R. Anstey and John A. Schuster (eds.), *The Science of Nature in the Seventeenth Century: Patterns of Change in Early Modern Natural Philosophy.* Dordrecht: Springer, pp. 9–33.

——— (2010). *How Modern Science Came Into the World.* Amsterdam: Amsterdam University Press.

Cunningham, Andrew and Perry Williams. "De-Centering the 'Big Picture': The Origins of Modern Science and the Modern Origins of Science," *The British Journal for the History of Science* (1993) 26(4): 407–432.

Dear, Peter. "Religion, Science and Natural Philosophy: Thoughts on Cunningham's Thesis," *Studies in History and Philosophy of Science* (2001) 32(2): 377–386.

——— "What Is the History of Science the History Of? Early Modern Roots of the Ideology of Modern Science," *Isis* (2005) 96(3): 390–406.

——— (2006). *The Intelligibility of Nature: How Science Makes Sense of the World.* Chicago: University of Chicago Press.

Endress, Gerhard (ed.). (2006). *Organizing Knowledge: Encyclopaedic Activities in the Pre-Eighteenth Century Islamic World.* Leiden: Brill.

Forman, Paul. "The Primacy of Science in Modernity, of Technology in Postmodernity, and of Ideology in the History of Technology," *History and Technology: An International Journal* (2007) 23: 1–152.

Gaukroger, Stephen (2006). *The Emergence of a Scientific Culture.* Cambridge: Cambridge University Press.

Gellner, Ernest (1964). *Thought and Change*. London: Weidenfeld and Nicolson.

Gerth, H.H. and C. Wright Mills (1958). *From Max Weber: Essays in Sociology*. New York: Oxford University Press.

Grant, Edward (1996). *The Foundations of Modern Science in the Middle Ages*. Cambridge: Cambridge University Press.

——— (2004). *Science and Religion, 400BC to 1550 AD*. Baltimore: Johns Hopkins.

——— (2007). *A History of Natural Philosophy: From the Ancient World to the Nineteenth Century*. Cambridge: Cambridge University Press.

Habermas, J. (1986). *The Theory of Communicative Action*. Volume 2. Cambridge: Polity Press.

Hall, A. Rupert (1954). *The Scientific Revolution, 1500–1800: The Formation of the Modern Scientific Attitude*. London: Longmans.

——— (1983). *The Revolution in Science: 1500–1750*. London: Longmans.

Henry, John (2008). *The Scientific Revolution and the Origins of Modern Science*. Basingstoke, Hampshire: Palgrave MacMillan.

Hogendijk, Jan P. and Abdelhamid I. Sabra (2003). *The Enterprise of Science in Islam: New Perspectives*. Cambridge, Massachusetts: The MIT Press.

Hoodbhoy, Pervez (1992). *Islam and Science: Religious Orthodoxy and the Battle for Rationality*. Kuala Lumpur: S. Abdul Majeed.

Huff, Toby (1995). *The Rise of Early Modern Science*. Cambridge: Cambridge University Press.

——— "Science and Metaphysics in the Three Religions of the Book," *Intellectual Discourse* (2000) 8(2): 173–198.

Kheirandish, Elaheh (2006). "Organizing Scientific Knowledge: The 'Mixed' Sciences in Early Classifications," in Gerhard Endress (ed.), *Organizing Knowledge: Encyclopædic Activities in the Pre-Eighteenth Century Islamic World*. Leiden: Brill, pp. 135–154.

Kuhn, Thomas S. (1970). *The Structure of Scientific Revolutions*. Chicago: Chicago University Press.

Lindberg, David C. (1992). *The Beginnings of Western Science*. Chicago: The University of Chicago Press.

Lindberg, David C. and Robert S. Westman (eds.) (1990). *Reappraisals of the Scientific Revolution*. Cambridge: Cambridge University Press.

Lloyd, Geoffrey. "The Comparative History of Pre-Modern Science: The Pitfalls and the Prizes," *Studies in History and Philosophy of Science* (1997) 28(2): 363–368.

Macfarlane, Alan (2007). "Ernest Gellner on Liberty and Modernity," in Sinisa Malesevic and Mark Haugaard (eds.), *Ernest Gellner and Contemporary Social Thought*. Cambridge: Cambridge University Press, pp. 31–49.

Mohd Hazim Shah. "Models, scientific realism, the intelligibility of nature, and their cultural significance," *Studies in History and Philosophy of Science* (2011) 42 (2): 253–261.

Nasr, S.H. (1968). *Science and Civilization in Islam*. Cambridge, Mass.: Harvard University Press.

———— (1976). *Islamic Science: An Illustrated Study*. London: World of Islam Festival.

———— (1988). "Islamic Science, Western Science: Common Heritage, Diverse Destinies," in Sardar (1988), pp. 250–259.

———— (1996). *Religion and the Order of Nature*. New York: Oxford University Press.

Park, Katharine and Lorraine Daston (eds.) (2006). *The Cambridge History of Science. Volume 3: Early Modern Science*. Cambridge: Cambridge University Press.

Rouse, Joseph. "What Are Cultural Studies of Scientific Knowledge?" *Configurations* (1993) 1(1): 57–94.

Sabra, A.I. "The Appropriation and Subsequent Naturalization of Greek Science in Medieval Islam: A Preliminary Statement," *History of Science* (1987) 25: 223–243.

———— "Science and Philosophy in Medieval Islamic Theology," *Zeitschrift fur Geschichte der Arabisch-Islamischen Wissenschaften* (1994) 9: 1–42.

———— "Situating Arabic Science: Locality versus Essence," *Isis* (1996) 87(4): 654–670.

———— (2003). "Ibn al-Haytham's Revolutionary Project in Optics: The Achievement and the Obstacle," in Jan P. Hogendijk and Abdelhamid I. Sabra (eds.), *The Enterprise of Science in Islam: New Perspectives*. Cambridge, Massachusetts: The MIT Press, pp. 85–118.

Sardar, Ziauddin (ed.) (1988). *The Revenge of Athena: Science, Exploitation and the Third World*. London and New York: Mansell.

Schneider, Mark A. (1993). *Culture and Enchantment*. Chicago: University of Chicago Press.

Shapin, Steven and Simon Schaffer (1985). *Leviathan and the Air Pump: Hobbes, Boyle and the Experimental Life*. Princeton, NJ: Princeton University Press.

Stenberg, Leif (1996). *The Islamization of Science: Four Muslim Positions Developing an Islamic Modernity*. Lund, Sweden: Lund University.

Westfall, Richard S. (1978). *The Construction of Modern Science: Mechanisms and Mechanics*. Cambridge: Cambridge University Press.

Zaidi, Ali Hassan. "Muslim Reconstructions of Knowledge and the Re-enchantment of Modernity," *Theory, Culture and Society* (2006) 23(5): 69–91.

CHAPTER 13

The Greatest Mistake: Teleology, Anthropomorphism, and the Rise of Science

Franklin Perkins

1 The Mutual Dependence of Anthropomorphism and Anthropocentrism

This essay takes the "comparativist" approach to the "Needham Question": Why did modern science emerge in Europe but not Asia, despite the greater achievements of Asian science through the previous millennium? What is most clever about Needham's question is that it cannot be answered through some essentialist account of "East" and "West." If the explanation lay in the different "essences" of European and Chinese cultures, then one would expect Europe to have been ahead all along. It is precisely this shift in the 16th and 17th centuries that suggests an explanation through the transmission of knowledge rather than cultural differences. Nonetheless, while the transmission of knowledge and technology from Asia provided a necessary condition for the development of modern science in Europe, the knowledge transmitted had to be taken up, applied, and theorized through shifts within European culture as well. I will here examine one of those shifts, contrasting it with the early development of Chinese thought.

Before turning to the main argument, it must be admitted that there is something ridiculous about even attempting to answer the Needham Question. At any given moment of history, there have been infinitely many differences between China and Europe. There is no way to distinguish which of those differences are relevant to explain the more rapid rise of science in Europe. Moreover, it seems impossible in principle to prove any answer. One has to show that something is true of *all* Chinese thinkers, but one can only examine specific cases. That is, it is not so difficult to explain why the philosopher, *Xunzi*, did not lead the way to modern science, but impossible to explain why *no one* did. That said, I will make an attempt, and while the answer cannot be fully adequate, I hope at least to illuminate real differences that have important consequences for the contrast between European modernity and the development of Chinese thought. The first part of the paper will examine Europe, with the second part turning to China.

© KONINKLIJKE BRILL NV, LEIDEN, 2016 | DOI 10.1163/9789004264199_015

"Modernity" is a vague concept that can and should be defined and characterized in many different ways, but a useful approach on the level of intellectual history is given by Susan Neiman in her book, *Evil in Modern Thought: An Alternative History of Philosophy*. According to Neiman, European modernity began and ended with two versions of the problem of evil. The first – symbolized by the Lisbon earthquake of 1755 – marked a final break from attempts to explain natural events by direct appeal to God's purposes. That shift started as early as Descartes's explicit rejection of final causes in the *Principles of Philosophy*, first published in 1644: "When dealing with natural things we will, then, never derive any explanations from the purposes which God or nature may have had in view when creating them."[1] The second event – which Neiman takes as marking the end of modernity and the start of "post-modernity" – is Auschwitz. Auschwitz symbolizes the final collapse of the Enlightenment's confidence in humanity and human progress. Neiman summarizes: "Lisbon revealed how remote the world is from the human; Auschwitz revealed the remoteness of humans from themselves" (Neiman 2002, 240).

These shifts are better seen as two stages in one prolonged break from anthropomorphic theism. The initial break, in which God's purposes were excluded from explaining the details of the natural world, did not suffice to dislodge the conception of human beings as *imago dei*, made in the image of God. That retained its grip on European thought for several more centuries (whether explicitly or implicitly). In fact, we still see traces of such a view whenever someone claims that human choices are free in a way that is radically different from the causality found everywhere else in nature, or when one assumes that the human mind is somehow uniquely commensurate with the structure of the universe. Understanding the function of this *imago dei* requires considering the mutual support between anthropomorphism and anthropocentrism. I take "anthropomorphism" in a broad sense to include any view which projects uniquely human characteristics into nature itself, so that it would include not just images of god as an old guy with a beard but also the Platonic view that the foundations of the world are divine ideas. I also take "anthropocentrism" broadly, as including any view which gives human values or knowledge a status in the world that is different in kind from those of other living things. It is likely that anthropomorphism arises as a projection of anthropocentrism – as Hume says, if spiders have a religion, their gods surely created the world by spinning a web (Hume 1998: 48). The relationship, though, is not unidirectional. Once anthropomorphic views are established, they defend anthropocentrism against the empirically plausible claim that we humans are merely, to use the

---

1 *Principles* I; translation from Cottingham et al. (1985), p. 28.

Chinese phrase, one of the tens of thousands of things (*wanwu* 萬物).[2] If we resemble God in some particular way, then we truly are special. Our values – what we label as good and bad – reflect the truth, unlike the values of fish, monkeys, or deer.[3] This is why it matters that we are made in the image of God, not vice versa. More importantly here, if the world is designed by a human-like God, then it is fundamentally commensurate with the human intellect. That is, however spiders understand the world, we assume that this knowledge is a limited view constructed from a spider perspective. Our knowledge is not like that, precisely because the world was designed according to concepts resembling ours. Such a view sets high expectations for human knowledge, making precise, comprehensive, systematic knowledge of the world into a plausible goal. Without it, one would likely be content with workable approximations that effectively promote human flourishing.

Heidegger illuminates this point with particular clarity, claiming that the conception of truth as the adequation of ideas to things [*veritas est adaequatio intellectus ad rem*] depends on a prior adequation of things to ideas [*adaequatio rei ad intellectum*]. He explains: "*Veritas* as *adaequatio rei ad intellectum* [...] implies the Christian theological belief that, with respect to what it is and whether it is, a matter, as created (*ens creatum*), *is* only insofar as it corresponds to the idea preconceived in the *intellectus divinus*, i.e., in the mind of God, and thus measures up to the idea (is correct) and in this sense is 'true.' "[4] That is, we can conceive of our ideas as matching things in the world only because of a more fundamental assumption that things themselves already match ideas. Of course, Heidegger is not claiming that this belief is always held consciously, but rather that the trust that human knowledge is precisely commensurable to the structure of the natural world is plausible only with some such assumption.

2   Science, Intelligibility, and the *Imago Dei*

It is likely impossible to prove that the rapid growth of science in modern Europe was a legacy of *imago dei*, but the connection was made explicitly by

---

2   The *Zhuangzi* states: "In counting things we say there are ten thousand, and human beings are just one of them. [...] In comparison with the ten-thousand things, [human beings] are not even like the tip of a hair to the body of a horse" (Guo 1978, 17: 564).

3   For a rejection of this view, see the dialogue between Gaptooth and Wang Ni in chapter two of the *Zhuangzi* (Guo 1978, 2: 93).

4   "On the Essence of Truth," translation from Krell (1993), p. 118.

intellectuals playing key roles in the scientific revolution. In fact, the dependence of modern science on anthropomorphic theism can be seen on several distinct levels. Most obviously, the fact that the world expressed God as its cause made the study of nature itself into a religious pursuit. Science reveals the glory of God. One finds this concern explicitly raised by many early modern scientists, from Kepler to Boyle to Newton to Leibniz. Consider one of Leibniz's justifications of science, appearing in a short essay entitled "Felicity":

> But one cannot love God without knowing his perfections, or his beauty. And since we can know him only in his emanations, there are two means of seeing his beauty, namely in the knowledge of eternal truths (which explain reasons in themselves) and in the knowledge of the Harmony of the Universe (in applying reasons to facts). That is to say, one must know the marvels of reason and the marvels of nature.... The marvels of physical nature are the system of the universe, the structure of the bodies of animals, the causes of the rainbow, of magnetism, of the ebb and flow, and a thousand other similar things. (Riley 1988: 84)

It is not quite accurate to say that the connection to God gave the investigation of nature an intrinsic value – it is still serving as a means – but it did give science a value aside from that of human utility. A second link between God and science relies on God's benevolence. The most famous example appears in Descartes's *Meditations on First Philosophy*, not so much in proving that this world is not a dream (which was not seriously doubted) but rather in guaranteeing that differences among our perceptions systematically map on to differences in the real world.[5] If that were not the case, we would be hopelessly deceived and scientific knowledge would be impossible. A good God would not put us in such a position.

In this essay, I want to focus on a third link, which is how the connection to God as an intelligent cause justified the intelligibility of nature. This claim is already implicit in the widespread description of the natural world as the Book of Nature, which implies a meaningful structure derived from an author. On a more abstract level, the applicability of our innate ideas (substance, cause and effect, mathematics) to the world follows because those ideas are shared

---

5 Descartes writes, "And from the fact that I perceive by my senses a great variety of colours, sounds, smells and tastes, as well as differences in heat, hardness and the like, I am correct in inferring that the bodies which are the source of these various sensory perceptions possess differences corresponding to them, though perhaps not resembling them" ("Sixth Meditation"; translation from Cottingham (1996), p. 56.

with the God who designed the world. This is the most direct application of the *imago dei*.[6] We can consider two explicit examples. The first is from a letter written by Johannes Kepler in 1599:

> Those laws [which govern the material world] lie within the power of understanding of the human mind; God wanted us to perceive them when he created us in His image in order that we may take part in His own thoughts... Our knowledge [of numbers and quantities] is of the same kind as God's, at least insofar as we can understand something of it in this mortal life.[7]

The second example comes from Wilhelm Gottfried Leibniz, from one of several "prefaces" he wrote for a "universal characteristic":

> There is an old saying that God made everything in accordance with weight, measure, and number. But there are things which cannot be weighed, namely, those that lack force and power, and there are also things that lack parts and thus cannot be measured. But there is nothing that cannot be numbered.[8]

The fact that the world was structured according to number gives Leibniz confidence in the possibility of an artificial language that would be able to represent all truths in a way susceptible to calculation, this "universal characteristic." Rather than argue endlessly about the existence of God, we would stop and say *calculemus* – Let us calculate! (Gerhardt 1978, VII, 200). Leibniz never discovered this universal characteristic (for which he hoped to make use of Chinese characters) but without this belief in its possibility, would he have made as much progress in developing formal logic? Would he have invented calculus?[9]

---

6 For a thorough discussion of this connection, see Jolley (1990).

7 Letter from Kepler to Herwart von Hohenburg, Apr. 9/10, 1599; translation from Holton (1988), p. 69.

8 "Preface to a Universal Characteristic (1678–79);" translation from Ariew and Garber (1989), p. 5.

9 This link between the intelligibility of nature and its origin in the mind of God was strongest among "rationalists" who appealed to innate ideas, but it appears widely in different forms. For example, drawing on manuscript materials, Stephen Snobelen concludes that for Newton, "God guarantees that both Scripture and nature can be understood by the human mind." [Snobelen (2001), p. 199, p. 202]. This appears specifically in the assumption that the natural world can be grasped through simple principles, because God acts with order and the greatest simplicity.

As is so often the case, Kant is remarkably astute in analyzing the tradition before him. Kant's whole epistemology can be seen as an attempt to maintain the earlier anthropocentrism of people like Descartes and Leibniz while admitting that its ground in anthropomorphism is unknowable. While (*contra* Descartes) there is no divine guarantee that our categories of substance or cause and effect fit reality itself, we cannot but take them as absolute. The most revealing point here is Kant's claim that the whole enterprise of science is possible only with the assumption of teleology, that is, with the assumption that this world is the product of a single intelligent source. Kant fully admits we cannot *know* that this is the case, but we must commit ourselves to it if we are to make sense of the world in a systematic way. Part of Kant's argument relies on our ability to grasp organic life, which he thinks cannot be explained without appeal to teleology (5: 375–76; 247).[10] Teleology is built into his very definition of organic life: "An organized product of nature is that in which everything is an end and reciprocally a means as well. Nothing in it is in vain, purposeless, or to be ascribed to a blind mechanism of nature" (5: 376–377; 248).

Once we admit that some things in nature must be explained teleologically, though, we are naturally and legitimately led to ask about the *telos* of nature as a whole (5: 398; 269). It is on the level of the whole that science requires teleology. To investigate empirical laws, the power of judgment must take it as an *a priori* principle that "in accordance with these laws a cognizable order of nature is possible." Kant explains:

> This agreement of nature with our faculty of cognition is presupposed *a priori* by the power of judgment in behalf of its reflection on nature in accordance with empirical laws, while at the same time the understanding recognizes it objectively as contingent, and only the power of judgment attributes it to nature as transcendental purposiveness (in relation to the cognitive faculty of the subject): because without presupposing this, we would have no order of nature in accordance with empirical laws, hence no guideline for an experience of this in all its multiplicity and for research into it. (5: 184–85; 71–72)

The laws of nature might be so diverse as to exceed human grasp, just as singular things might be ultimately irreducible to a manageable set of genera and species. As Kant says, nature might be "only infinitely manifold and not fitted

---

10   For citations of Kant, I have used the translation in Guyer and Matthews (2000). Citations are by volume and page number in *Kants gesammelte Schriften* (Kant 1900–), followed by page number in Guyer and Matthews (2000).

for our power of comprehension" (5: 185; 72). Such an assumption, though, would make it impossible to even begin the systematic investigation of nature. Science requires the assumption that nature has a comprehensible order reflecting its origin in another mind, the mind of God (5: 407; 276). Kant is, of course, not claiming that those pursuing science consciously appeal to teleology, but they use principles that rely on it. He lists examples such as, "Nature takes the shortest way" or "the great multiplicity of its empirical laws is nevertheless unity under a few principles" (5: 182; 69).[11] In general terms, Kant's point is that the pursuit of scientific knowledge requires the belief that the world itself is not so diverse and chaotic as to exceed the capacities of human judgment, which then requires the belief that nature was rationally organized by a single creator.[12]

In a global context, views that explain events in the natural world by appeal to the intentions of anthropomorphic divinities are of course quite common, ranging from more abstract claims for intelligent design to explanations of rain and drought in terms of spirits. A view that rejects both anthropomorphism and anthropocentrism – one that takes human beings as merely one of the myriad things – also is coherent and, if not common, at least present in different times and places. Using the trajectory of European history, we might call the first views "pre-modern" and the second "post-modern." What is peculiar is the middle ground of European modernity, which attempted to break from theistic explanations while still holding onto human beings as godlike. The afterimage of God remained long after God himself faded into the background. The tension inherent in this view appears throughout the modern period, where thinkers rejected appeals to final causes while resting their epistemological foundations on God.[13] Descartes invokes God within his very argument against appeals to divine intentions, saying we should instead use the natural light God has planted in us. Kant sees the problem clearly, taking great care to explain how the antinomy between mechanistic causality and teleology can be avoided. If we inquire into the causes of this lag time between the decline of God and the decline of the image of God, we run into the same problems of any causal explanation of historical changes. Nonetheless, one explanation is fairly

---

11   Leibniz makes the same point in section 21 of the *Discourse on Metaphysics* [Ariew and Garber (1989), pp. 53–55].

12   My use of Kant here is indebted to discussions with Avery Goldman and to the book Goldman (2012).

13   In fact, this tension appears within discussions of final causes as well, where immanent final causes were rejected in explaining particular things at the same time that nature as a whole was seen as designed by God. On this tension, see Osler (2001).

obvious – in Europe, an account of the world in terms of an anthropomorphic god was enforced through the threat of violence for well over a thousand years. Such control over theoretical inquiry is, I think, unprecedented. One result is that by the time philosophers were able to question the role of God, God had already been thoroughly incorporated into every aspect of European thought. Under such conditions, it is unsurprising that it took several centuries and the horrors of the 20th century for philosophers to fully come to terms with the "Death of God."

3   The Origins of Chinese "Humanism"

The mutual implication of anthropomorphism and anthropocentrism did appear in China, and it is usually traced back to the doctrine of *tianming* 天命, the "Mandate of Heaven," which was used to rationalize the Zhou conquest of the Shang, around the 12th century BCE.[14] According to this view, heaven cared for the people, so it would punish rulers who harmed the people and elevate good leaders in their place. While this view is more limited than Christian claims about God – there is no evidence that the early Zhou people believed that heaven was omnipotent or that good people were always rewarded – it still provided a relatively secure footing for human values by grounding them in a humanlike divinity. This shift is commonly given as the origin of so-called Chinese "humanism" (*renwen zhuyi* 人文主義). For example, Wing-Tsit Chan and Xu Fuguan 徐復觀 both begin their influential accounts of the history of Chinese philosophy by appeal to the humanism that resulted from belief in the "Mandate of Heaven."[15] Xu Fuguan explains:

> Thereupon, the mandate of heaven (divine intent) no longer was the unconditional support for certain rulers but made selections based on human actions. In this way, the mandate of heaven gradually moved out from a spirit of mystery and darkness and became something that could be understood and grasped through human beings' own actions.

---

14   Many of the points in this section are discussed in more detail in Perkins (2014), which focuses on the role of the problem of evil in the formation of classical Chinese philosophy.

15   The first section of Chan's highly influential *Sourcebook of Chinese Philosophy* is titled "The Rise of Chinese Humanism." Chan (1963). Xu characterizes the Zhou dynasty as a progression toward a "humanistic spirit" (*renwen jingshen* 人文精神). Xu (1969), p. 15.

Moreover, it became the ultimate guarantee for the rationality of human actions (Xu 1969: 24).[16]

The contemporary Chinese philosopher Chen Lai similarly claims that the fall of the Shang led the Zhou to the following realization:

> Human beings cannot attribute all worldly events to the necessity of the heaven's commands. History is not entirely determined by heaven (Shang Di) but human actions really participate in historical processes, so that human beings should seek for the causality of historical change within human action itself (Chen Lai 2009: 191).

With the doctrine of the Mandate of Heaven, responsibility for order shifts to human beings. As Xu Fuguan says, it allowed a move from reliance on spirits to trust in the power of one's own actions (Xu 1969: 21). This humanistic trust, though, follows from the belief that heaven itself shares human values and human concerns.

The collapse of the Zhou Dynasty and the descent of China into centuries of war and chaos undermined the belief in the goodness of heaven and the corresponding confidence in human power. Reflecting back on that history, Sima Qian famously concludes his discussion of the death by starvation of the virtuous brothers Bo Yi and Shu Qi with these words:

> When it comes to the present age, there are those acting recklessly off the track, focused only on violating taboos, but they live out their lives in ease and joy, enjoying abundant prosperity lasting for generations without being cut off. Others choose their ground carefully and tread there, speaking only in a timely way, their actions not following deviant paths, indignant only at what is not fair or correct, but they meet misfortune and disaster. Such people are more than can be counted. So I am deeply perplexed by this. What we call the way of heaven – is it right? is it wrong? (Sima Qian 1959: 61.2125)

As with the beginning and end of "modernity" in Europe, the key transformative event in early China can be seen as an encounter with the "problem of evil." The sad reality of Chinese history in the Spring and Autumn and Warring States Periods made it almost impossible to maintain a view of heaven as good. Many views of the divine remained through the Spring and Autumn

---

16   All translations from Chinese are my own.

and Warring States periods, but the dominant discourses gave up the view that heaven is consistently good and refused to explain natural phenomena in terms of the intentions of spirits.

This shift bears undeniable analogies with the start of the Enlightenment in Europe and the break from the "pre-modern." One finds, for example, critiques of superstition that would easily fit in 18th century Europe. A nice example was recently discovered on a 4th century BCE bamboo strip. In this short dialogue between Kongzi and his disciple Zigong, the power of sacrifices to mountains and rivers to bring rain is rejected with the following argument:

> Now mountains have stones for skin and trees for people. If heaven does not rain, the stones will roast and the trees will die. The mountains desire rain more than us – why would they wait for us to call on them! Now rivers have water for skin and fish for people. If heaven does not rain, the water will dry up and the fish will die. The rivers desire water more than us – why would they wait for us to call on them! (Ji Xusheng 2003, strips 4–5).[17]

Chen Jialing 陳嘉凌, one of the editors of the bamboo text, says that the dialogue shows Zigong's "attitude of rationalist reform," which he takes as projecting toward the views of Xunzi (Ji Xusheng 2003: 41).

A similar critical attitude is developed further in the 3rd century BCE by Xunzi, as in his approach to bad omens:

> Falling stars and strange calls are things people and states all fear. For what reason do they happen? There is no reason. They are the changes of heaven and earth, the transformations of *yin* and *yang*, things which rarely arrive. Considering them strange is acceptable, but fearing them is wrong. Now eclipses of the sun and moon, untimeliness of wind and rain, strange stars appearing together – these are things which every age has had. If those above are insightful and the government is balanced, then even if these all arise together there will be no harm. If those above are darkened and the government is crooked, then even without one of these arriving, there will be no benefit. (17: 313–314)[18]

---

17   The bamboo text has been published in the Shanghai Museum collection under the title, "Great Drought in the State of Lu" (魯邦之旱). While the structure of the text is unclear, it seems that these words are spoken by Zigong. Versions of the same conversation appear in other texts; for a discussion of these connections, see Cao (2006), pp. 93–106.

18   Citations of the *Xunzi* are by chapter and page number in Wang (1988).

The passage continues by arguing that what should be feared as ill omens are human actions: disordered government, oppression of the people, disrupting the agricultural seasons (17: 313–314). In another passage, Xunzi opposes claims that rituals and sacrifices influence heaven or other spirits, claiming that it is "inauspicious" to believe that the rituals are done for the sake of spirits (17: 316). This critical attitude of "de-mystification" also appears in Xunzi's rejection of prognostication and his co-opting of the terms associated with it.

Xunzi not only rejects appeals to spirits and omens but opposes any role for the will of heaven:

> The course [*xing* 行, actions] of heaven has regularity: it is not that it exists for a Yao and does not exist for a Jie. Respond to it with order and it will be propitious. Respond to it with disorder and it will be unpropitious... If you cultivate the way and do not err, then heaven cannot give misfortune. Thus floods and droughts cannot cause starvation, cold and heat cannot cause sickness, and omens and aberrations cannot be unpropitious.... If the way is violated and actions are reckless, then heaven cannot make it propitious. Thus even without floods and droughts arriving, there will be starvation, without cold and heat oppressing, there will be sickness, and without omens and aberrations arriving, it will be unpropitious.... Thus one who understands the division between heaven and human can be called a person who has reached the utmost. (17: 306–308)

The point of this passage is to emphasize that the responsibility for order and disorder lies entirely with human beings, but it is remarkably defiant. If you do the right things (i.e., actions which are causally linked to prosperity, order, and honor), then heaven has no power to harm you. Another passage says that the causal power of correct action "is something heaven cannot kill, earth cannot bury, and an age like that of [the evil emperor] Jie or Robber Zhi cannot pollute" (8: 139). This view is rooted in natural causality. In a passage explaining why virtue brings success, Xunzi writes:

> The arising of any kind of thing necessarily has something from which it begins. The arrival of glory or shame necessarily is an image of one's virtue (*de* 德). Rotten meat gives out maggots, decaying fish give birth to worms. With negligence and forgetting one's person, harm and disaster appear. (Wang 1: 6–7)

Bad actions bring harm and shame, but they are linked by the same causal necessity that applies everywhere else in nature. Heaven and spirits have no influence.

This declaration of independence from divine intentions has remarkable similarities with the move that Neiman takes as marking the start of European modernity. The very choice by Chinese scholars to label it as a "humanistic turn" is meant to invoke this analogy. Seeing that we could not in practice rely on the goodness of God or even understand the world in terms of God's specific purposes, responsibility was seen to shift to human beings themselves.[19] This shift in responsibility is precisely what motivates Xunzi's claims. Knowing the difference between heaven and human (*tianrenzhifen* 天人之分) means recognizing that the power to control our lives is with us, not heaven. The similarities between Xunzi and thinkers of the European Enlightenment has been noted and emphasized by Chinese scholars seeking sources for scientific thought within their own traditions. For example, Hu Shi links Xunzi's position with Francis Bacon's "Conquest of Nature" (Hu 2003: 239).[20] Chen Daqi also connects Xunzi to Bacon, explaining:

> This passage from Xunzi can be said to be his most splendid statement for the natural sciences and also most worthy of the attention of later generations. Xunzi wants to make things and nature serve and wants to increase human power to increase natural production. This is not unaligned with the direction of Westerners toward conquering nature and is very close to the spirit of modern natural sciences. (Chen Daqi 1954: 21)

Chen goes on to lament that Xunzi was marginalized in the Chinese tradition, claiming that if this had not been the case, science in China would not have fallen so far behind. In other words, Chen Daqi's answer to Needham's question is that the slow progress of science in China was not due to a lack of indigenous resources but to the fact that those resources were ignored.

4   The Limits of the Human

Chen Daqi is wrong, though, at least about Xunzi. While there might be some truth in attributing to Xunzi something like a scientific inquiry into human action (some mix of sociology and psychology), there is no evidence of any trajectory toward the scientific study of nature. For that, we would have to look

---

19   Neiman says: "Modern conceptions of evil were developed in the attempt to stop blaming God for the state of the world, and to take responsibility for it on our own." Neiman (2002), p. 4.

20   Hu Shi himself provides the English phrase as a gloss to the Chinese phrase *kantian zhuyi* 戡天主義.

instead to the Mohists, who did leave systematic definitions of terms relating to science, mathematics, technology, and logic. What is striking, though, is that the Mohists break much less clearly from the "pre-modern" view that relied on divine intentions. They rely explicitly on the "will of heaven" both as an enforcer of right and wrong and as a standard. As the *Mozi* says:

> I have heaven's intention like wheelwrights have the compass and wood-workers have the square. The wheelwright and woodworker hold their compass and square in order to measure what is rectangular and circular in the world, saying: "what fits is right, what doesn't fit is wrong." Now the writings of the world's scholar-officials and gentlemen are more than can be listed, and their sayings are more than can be counted. Above they persuade feudal lords, below they persuade outstanding scholar-officials. Yet in their rightness and humaneness, they are very far apart. How do I know this? I say: I have attained the world's clearest standard to measure them. (Sun Yirang 2001, 26: 197)

It might initially strike us as odd or ironic that the Warring States school with the most confidence in human knowledge was also exceptional in holding on to appeals to divine intentions (including even ghosts), but given the inter-dependence of anthropomorphism and anthropocentrism, it is just what we would expect.[21]

If we take the Mohists as one side in maintaining anthropomorphism and confidence in the human, Xunzi is on the other side.[22] He is in some ways most like European thinkers who broke from appeals to divine intentions, but also least like them. We now have a framework for explaining this peculiarity (a peculiarity more with Europe than Xunzi). In the terms I have been using, we could say that Xunzi moves directly from a pre-modern to a post-modern view, skipping the lag time between the collapse of the divine and confidence in human power.[23] We can also see an approach to the Needham Question, which is not why China did not go far in science but rather the paradoxical fact

21   In fact, the Mohists are still relatively skeptical in comparison to most thinkers of the European Enlightenment, and the support that heaven gives to human knowledge is much weaker than that given by God for thinkers like Descartes or Leibniz. A discussion of this point is beyond the scope of this paper, but see Perkins (2011), pp. 76–78.

22   The *Zhuangzi* goes even further in that direction, but the *Zhuangzi* is so far out that even a contrast with a scientific viewpoint is difficult to make.

23   My use of "pre-modern" and "post-modern" is simply for orientation and should not be taken as implying a wish to universalize historical periods derived from Europe. On the contrary, my hope is to undermine such universalizations by showing how they become scrambling when applied outside of Europe.

that China was ahead in science and technology and then was surpassed. The naturalistic "postmodern" view we find in Xunzi is surely more conducive to science and technology than earlier views that would explain natural events in terms of either the will of heaven or the activities of spirits. It thus has clear advantages over the views that dominated through most of European history. Such a view, though, is not as favorable toward the development of science as one which assumes, to use Kant's phrase, the "agreement of nature with our faculty of cognition."

To move past mere assertions, I will briefly consider three specific points from the *Xunzi* – the limited and perspectival nature of human knowledge, the need to restrict inquiry to the human realm, and the criticism of individual judgment. All three contrast the assumptions of most Enlightenment philosophers. For Xunzi, the origin of knowledge is in sensory perception, which involves awareness of multiplicity and difference. Based on similarities and differences, we formulate names and categories in order to organize things (22: 415–416). Higher levels of knowledge involve the ability to apply these categories and make inferences. The formation of knowledge, though, is far more difficult than this suggests. The problem is that experience can be categorized in infinitely many ways, making it profoundly difficult to develop a stable and effective system of names and categories. Xunzi explains this in a passage that discusses both the possibility of knowledge and its limits:

> In general, that by which one knows is human nature [性 *xing*], and that which can be known is the coherent patterns [理 *li*] of things. With a human nature that can know, one seeks the coherent patterns of things that can be known. Yet if you have nothing by which to stop and limit it, then even in exhausting your years and going to the limit of your life, you cannot get through it all. (21: 406–407)

We can know the world because we naturally have an ability to learn and the world has knowable patterns. This match between the world and our abilities, though, is ultimately a mismatch, since our capabilities are limited while the patterns that can be known are not. For Xunzi, the problem is not in knowing the world, but rather in reconciling that knowledge with the infinite complexity of nature itself. In this sense, Xunzi would likely accept the view Kant admits as a possibility but rejects as practically unthinkable: that nature is "infinitely manifold and not fitted to our power of comprehension."

This leads into the second point. Because the nature of reality so far exceeds the cognitive capabilities of human beings, we must be careful about where we direct our attention. Given the infinity of the world, we could progress forever without getting anywhere useful. We must restrict our investigations to topics directly useful for human life. We see the consequences of this orientation in a famous passage that perfectly unites the way Xunzi does and does not seem like an Enlightenment thinker:

> Magnifying heaven and thinking of it longingly – how can this compare with raising things and arranging them?
> Following heaven and singing its praise – how can this compare with arranging what heaven mandates [*tianming*] and using it? ...
> To yearn for that by which things are generated – how can this compare with having that by which things are completed?
> Thus to discard the human and think longingly of heaven is to lose the genuine characteristics of the ten thousand things. (17: 317)

Xunzi here advocates an orientation toward using the resources that the world provides and doing so in a thoughtful critical way, without speculation on divine intentions or reliance on divine help. In setting aside concerns with the divine, though, Xunzi simultaneously sets aside broader questions that lie at the foundations of science, questions like how things are generated. A similar attitude appears in Xunzi's dismissal of eclipses and natural aberrations as something that should be ignored rather than investigated. Thus Henry Rosemont Jr. says that Xunzi's view created "an intellectual atmosphere that was inimical to the conduct of pure scientific inquiry" (Rosemont 2000, 13). Antonio Cua also takes this passage as turning away from causal investigations into nature and thus as harmful to the development of science (Cua 1985: 27).[24]

We can now turn to the third point. Nature so far exceeds our cognitive capacities that even if we restrict ourselves to what is directly useful, we can never get very far on our own. Our situation, though, is not hopeless. The incommensurability between a single human being and the infinity of nature is bridged if we shift to a large human community working together over thousands of years. Over time, human beings develop a system of names and practices that at least provide a reliable way through the world. This way, though, is beyond the grasp of any one individual, at least in the beginning. We simply submit to it and follow it, at best coming to understand its grounding after years of practice and training. Thus while Mozi uses the will of heaven as his "compass and square," Xunzi tells us we must use the way of the sagely kings as our measure (19: 536). Several passages present ritual and tradition as the *only* guide:

---

24   Hu Shi makes a similar criticism of Xunzi. Hu (2003), p. 239.

Thus to negate the rituals is to be without a model; to negate the teacher is to be without a teacher. To not affirm teachers and models but love to use oneself instead – this is like having a blind person distinguish colors or a deaf person distinguish sounds. (2: 33–34)

Xunzi repeatedly criticizes those who innovate on their own, and in one infamous passage, Xunzi explains that if gentlemen had the power they would eliminate those with wicked doctrines. Lacking power, they are forced to debate and persuade (22: 422, 6: 98–99). In another passage, he characterizes misuse of names as a crime like falsifying tallies and measurements (22: 414). This authoritarianism is not accidental. It follows from doubts that the a lone individual could ever comprehensively figure out the complexity of the natural world itself, doubts that are quite reasonable once we give up the faith that human beings were made in the image of God. Such a view, though, is much less conducive to the development of science.

5   Conclusion: The Greatest Mistake?

Let me conclude first by summarizing what I hope to have shown. The unusually long and thorough control that religious authorities exerted over intellectual discourse in Europe set up a peculiar circumstance in which philosophy developed to a high level of sophistication while still remaining thoroughly based in an anthropomorphic theism. As a result, even when that theism broke down, the status of human beings which it had justified, human beings as the image of God, remained. This period of remainder marks the Enlightenment and European modernity, a period which broke down only in the 20th century. It was in this period, when explanations by appeal to divine intentions faded into the background without undermining the belief that the human mind was uniquely suited to grasp the intelligible order of the world, that science entered a period of rapid progress. This reply to the Needham Question is just the opposite of two common yet baseless accounts of why science rose in Europe – greater freedom of thought and a cleaner split from religion. While China had (and continued to have) analogous views of anthropomorphic spirits, such views were breaking down at the time Chinese philosophy first began to appear. Thus almost from the start Chinese philosophy was pervaded by a much more humble view of human beings as merely one of the myriad things in nature.[25] While this was initially an advantage in understanding the

---

25   It was common for Chinese philosophers (particularly the Ru) to consider human beings as the most noble of all things, but even in that context, human beings differ by degree

phenomena of the natural world, it never generated the confidence in human judgment required for the rise of modern science.

Let me conclude with one final irony. Even if some people would resist the claim that we have now entered a "post" modern period, I think most philosophers and scientists would reject the human exceptionalism at the foundations of the Enlightenment. Most are now much closer to the Chinese view of human beings as one of the myriad things. Science itself has led to this conclusion. Kant boldly asserts that it would be "absurd" for human beings to expect that there will ever be a "Newton" who can explain how even a single blade grass could arise through mechanistic causes (5: 400; 271). Darwin was that Newton. If Kant was right that the development of the modern form of science required the assumption of teleology and design, then science itself eventually overturned the very beliefs that made the rise of (modern) science possible. Another way to put it is that the rapid rise of modern science and technology in Europe may have required something we would now consider a mistake, a mistake not made in China, but a mistake with very great consequences. One might even call it, the greatest mistake. If that is indeed the case, then the views that led to the development of modern science may not be the best foundation for understanding what science actually is and how it works. While the more humble views of knowledge found in the Chinese tradition may not have produced a "scientific revolution," they may still provide insights into the status of science and the relationship between human beings and nature.[26]

Bibliography

Ariew, Roger, and Daniel Garber (eds. and trans.) (1989). *Leibniz: Philosophical Essays*. Indianapolis: Hackett.
Cao Feng 曹峰 (2006). *Shanghai chujian sixiang yanjiu* 上海楚簡想研究. Taibei: Wanjuanlou.

---

rather than by kind. Similarly, the systems of correlations that developed in the Han Dynasty allowed immense influence for human beings over nature, but this again was a matter of degree. Under such a system of "correlative cosmology," all things influence each other.

26   Versions of the paper were presented at the Asian Research Institute of the National University of Singapore and in the philosophy department at Nanyang Technological University, greatly benefitting from those discussions. I am particularly grateful for feedback and assistance from Roger Ames, Lina Jansson, and Brook Ziporyn.

Chan, Wing-tsit (1963). *A Sourcebook in Chinese Philosophy*. Princeton: Princeton University Press.

Chen Daqi 陳大齊 (1954). *Xunzi xueshuo* 荀子學説 (The Doctrines of Xunzi). Taibei: Zhonghua Wenhua Chuban Shiye Weiyuanhui Chubanshe.

Chen Lai 陳來 (2009). *Gudai zongjiao yu lunli: Rujia sixiang de genyuan* 古代宗教與倫理：儒家思想的根源 (Ancient Religion and Ethics: The Roots of Confucian Thought). Beijing: SDX Joint Publishing.

Chen Qiyou 陳奇猷 (1984). *Lüshi Chunqiu xinshi* 呂氏春秋新釋 (New Explanation of the *Lüshi Chunqiu*). Shanghai: Shanghai Guji Chubanshe.

Cottingham, John (ed.) (1996). *Descartes: Meditations on First Philosophy, with Selections from the Objections and Replies*. Cambridge: Cambridge University Press.

Cottingham, John, Robert Stoothoff, and Dugold Murdoch (eds. & trans.) (1985). *The Philosophical Writings of Descartes*. Vol II. Cambridge: Cambridge University Press.

Cua, Antonio S. (1985). *Ethical Argumentation: A Study in Hsün Tzu's Moral Epistemology*. Honolulu: University of Hawaii Press.

Gerhardt, C. J. (ed.) (1978). *Die Philosophischen Schriften von Gottfried Wilhelm Leibniz*. Hildesheim: George Olms Verlag.

Goldman, Avery (2012). *Kant and the Subject of Critique: On the Regulative Role of the Psychological Idea*. Bloomington: Indiana University Press.

Guo Qingfan 郭慶藩 (1978). *Zhuangzi jishi* 莊子集釋 (Collected Explanations of the *Zhuangzi*). Beijing: Zhonghua Shuju.

Guyer, Paul, and Eric Matthews (trans.) (2000). Immanuel Kant's *Critique of the Power of Judgment*. Cambridge: Cambridge University Press.

Holton, Gerald (1988). *Thematic Origins of Scientific Thought: Kepler to Einstein* (revised edition). Cambridge: Harvard University Press.

Hu Shi 胡适 (2003). *Zhongguo zhexueshi dagang* 中國哲學史大綱 (Outline of Chinese Philosophy). Beijing: Dongfang Chubanshe.

Hume, David (1998). *Dialogues Concerning Natural Religion*. Richard H. Popkin (ed.), 2nd edition. Indianapolis: Hackett Publishing.

Ji Xusheng, 季旭昇 (ed.) (2003). *Shanghai Bowuguan cang Zhanguo Chuzhujian II duben* 上海博物館藏戰國楚竹簡 (II) 讀本 (Shanghai Museum Warring States Chu Bamboo Strips Reader II). Taibei: Wanjuanlou Tushu Fufen.

Jolley, Nicholas (1990). *The Light of the Soul: Theories of Ideas in Leibniz, Malebranche, and Descartes*. Oxford: Oxford University Press.

Kant, Immanuel (1900–). *Kants gesammelte Schriften*. Berlin: German Academy of Sciences.

Krell, David (trans.) (1993). *Heidegger: Basic Writings*. Bloomington: Indiana University Press.

Neiman, Susan (2002). *Evil in Modern Thought: An Alternative History of Philosophy*. Princeton: Princeton University Press.

Osler, Margaret J. "Whose Ends? Teleology in Early Modern Philosophy," *Osiris* (2001) 16: 151–168.

Perkins, Franklin (2011). "No Need for Hemlock: Mengzi's Defense of Tradition," in Chris Fraser, Dan Robins, and Timothy O'Leary (eds.), *Ethics in Early China*. Hong Kong: University of Hong Kong Press, pp. 65–81.

——— (2014). *Heaven and Earth Are Not Humane: The Problem of Evil in Early Chinese Philosophy*. Bloomington: Indiana University Press.

Riley, Patrick (1988). *Leibniz: Political Writings*. Cambridge: Cambridge University Press.

Rosemont, Henry Jr. (2000). "State and Society in the Xunzi: A Philosophical Commentary," in T.C. Kline and Philip J. Ivanhoe (eds.), *Virtue, Nature, and Moral Agency in the Xunzi*. Indianapolis: Hackett, 1–38.

Sima Qian 司馬遷. (1959). *Shi ji* 史記 (Records of the Grand Historian). Bejing: Zhonghua Shuju.

Snobelen, Stephen D. " 'God of Gods, and Lord of Lords': The Theology of Isaac Newton's General Scholium to the Principia," *Osiris* (2001) 16: 169–208.

Sun Yirang 孫詒讓 (2001). *Mozi xiangu* 墨子閒詁 (Casual Notes on the *Mozi*). Beijing: Zhonghua Shuju.

Wang Xianqian 王先謙 (1988). *Xunzi jijie* 荀子集解 (Collected Explanations of the *Xunzi*). Beijing: Zhonghua Shuju.

Xu Fuguan 徐復觀 (1969). *Zhongguo renxinglun shi* 中國人性論史 (A History of Chinese Theories of Human Nature). Taibei: Taiwan Shangwu Yinshuguan.

CHAPTER 14

Rescuing Science from Civilisation: On Joseph Needham's "Asiatic Mode of (Knowledge) Production"*

Kapil Raj

1 Introduction

If there is one belief capable of drumming up a large consensus amongst historians of science, it is surely about the western European origins of modern science, most commonly situated in the Scientific Revolution of the 16th and 17th centuries and held to owe nothing to other cultures or times. The reasons for its putative emergence within the narrow boundaries of western Europe has been the subject of a plethora of writings celebrating the epistemological, sociological, and economic uniqueness of the West which have continued to appear ever since the emergence of the history of science as a full-fledged discipline in the early 20th century. Indeed, of all the questions dealt with by the history of science, this is probably the one for which the discipline is generally best known.[1]

It is in this sea of hubris that Joseph Needham's was a rare voice which did more than just assert the uniqueness and superiority of the West in the creation of modern science. Drawing on a staggering quantity of sources in Chinese, Japanese and European languages, Needham set himself a transcontinental comparativist agenda to identify precisely what constituted this putative uniqueness of western Europe. Like many intellectuals of his generation,

[handwritten marginal note: but not in this book]

---

* With apologies to Prasenjit Duara for taking liberties with the title of his seminal book, *Rescuing History from the Nation: Questioning Narratives of Modern China* (Chicago: University of Chicago Press, 1996). *Acknowledgements*: I wish to thank Luc Berlivet, Anne Cheng, Snait Gissis, Eivind Kahrs, Chien-Ling Liu, Mike Osborne, Simon Schaffer, Sam Schweber, Sanjay Subrahmanyam and the editors of this volume for their invaluable remarks and suggestions.

1 Although European autarky has always underpinned history of science writing – see for instance Duhem (1913–1959) – the *locus classicus* for this question is Butterfield (1949). See also Zilsel (2003); Koyré (1958, 1968); Hall (1954) *The Scientific Revolution 1500–1800: The Formation of the Modern Scientific Attitude*, published in its 2nd edition as *The Revolution in Science* (1983); and Westfall (1992). For a critical appraisal of this quest for origins, see Cunningham & Williams (1993).

Needham was convinced of the universality of science as a human enterprise, the expression of an innate curiosity fundamental to human nature throughout time and space.[2] Intrigued by the momentous scientific and technological achievements of China until the 15th century, he asked why modern science did not rise there, but originated only in Europe. The answer to what has come to be called Needham's "Grand Question" lay, according to him, in the resilience of the Confucian culture of China's agrarian bureaucracy which hindered the emergence of mercantile and industrial capitalism, a *sine qua non* in his view for the emergence of mathematical rationality, the bedrock on which modern science stood. Chinese, like Indian, or Arab science was based on local "ethnic-bound" categories which allowed the diffusion of technical innovations, but prevented that of their underlying theoretical systems. On the other hand, because it was founded on mathematical reasoning, modern science could be appropriated by all humans without any difficulty, and was thus "ecumenical". Yet, despite its uniqueness, modern science was not created *ex nihilo*. Rather, it subsumed the medieval learning of both West and East, "like rivers flowing into the ocean of modern science."[3] For Needham, then, while modern science originated in the West, yet it is culturally universal.

Ever since he formulated it almost half a century ago, Needham's "Grand Question" has constituted one of the major determinants in shaping global comparative approaches to the history of science.[4] Along with its Weberian counterpart ("Why did industrial capitalism not emerge in China?"), this question is, indeed, at the core of the "divergence" approaches in world history of a number of prominent historians.[5] However, unlike Needham who used the yardstick of scientific and technological dynamism to calibrate the protagonists, the latter base their comparison mainly on quantifiable data series and use the more economically based marker of the "Industrial Revolution" to differentiate between societies.

Another difference between the two approaches lies in the nature of the supposed factors used to explain non-occurrence. Needham sought his answers

---

2   For the Zeitgeist of Needham's generation, see Werskey (1988).

3   Needham (1970), p. 3, on p. 397. Needham never did arrive at a definitive answer to his "Grand Question", see his *Science and Civilisation in China*, 7 volumes (1954–2005) (henceforth SCC), especially Vol. 7, pt. 2, most of which is devoted to his various approaches to the question. For a critical evaluation of Needham's theses, see Sivin (1995).

4   See, for example, Habib & Raina (1999).

5   For Max Weber on the "Why not China?" question, see his "Konfuzianismus und Taoismus"(1915), translated into English by Gerth as *The Religion of China: Confucianism and Taoism* (1951). Among the best known relevant historical works are Wong (1997); Pomeranz (2000). See also Rosenthal and Wong (2011); and Parthasarathy (2011).

in essentialist structural features of the societies being compared, while the "divergence theorists", with the notable exception of the inventor of the term, Samuel Huntington, couch theirs in contingent historical factors. Indeed, the title itself of the former's magnum opus – *Science and Civilisation in China* – is a clear announcement of this approach, thus putting him squarely in the camp of "civilisationalism" (if you'll pardon this neologism), a major presupposition that continues, it must be said, to be widely held in the social sciences.[6] Looking for his explanations in the social, intellectual and economic structures of each civilisation, Needham was greatly influenced by Karl Wittfogel's concept of "bureaucratic feudalism" which itself was derived from Marx and Engels' scattered reflections on the "Asiatic mode of production".[7] Needham argued that as a result of their obsession with statecraft, ethical conduct and human affairs, China's despotic scholar-élite redistributed taxes in gigantic hydraulic-based public works, mainly river control and irrigation and transport canals, but had no taste for knowing the natural world. "The *Philosophia perennis* of China," he wrote, "was an organic materialism [...] the mechanical view of the world simply did not develop in Chinese thought."[8] On the other hand, the "divergence theorists" have resolutely turned their back to this sort of essentialist reasoning, seeking instead to situate the moment of disparity between East and West around the turn of the 19th century, with some even viewing the relative advantage of the West as being conjunctural and momentary.[9]

Notwithstanding these important differences, both variants of the "divergence" approach are open to the serious and fundamental critique – pertaining to the underlying counterfactual method – that they only seek to identify missing features and shortcomings of the non-West as seen through the prism of putative unique characteristics that they identify in the West; neither is interested in investigating the inherent dynamics of these "Other" regions.[10] Concerning science and technology in particular, Needham like all his

---

6   On Needham's deep-seated notion that human societies are organised according to civilisations, see his "Poverties and Triumphs of the Chinese Scientific Tradition", in Needham (1969), pp. 14–54, on pp. 20–23. For a critical historiography of the concept of civilisation, see Mazlish (2004). See also Subrahmanyam (2001, 2013); and Thapar (1999). On the Asiatic mode of production, see Godelier (1964).

7   See Wittfogel (1931). For Needham's debt to Wittfogel, see his "Science and Society in East and West", in *The Grand Titration*, pp. 190–217, on pp. 193 et seq., and *SCC*, v. 7, pt. 2, pp. 13–14.

8   Needham, "Poverties and Triumphs", p. 21.

9   For a typical example of non-essentialist reasoning see Elvin (1984).

10   For critiques of the counterfactual strategy, see especially Sivin (1982, 1983, 1985). See also Hamilton (1985).

contemporaries conceived of science as a unique means of knowing nature through a series of discoveries. Indeed, his "grand titration" method "to find out where credit is due" was based on a catalogue of discoveries attributed to different civilisations and the subsequent debates on Needham's thesis have all shared the same conception of pre-modern science as a system of propositions or discoveries derived from "ethnically-bound concept systems" as opposed to modern science, which albeit seen in terms of discoveries is based on "the mathematisation of hypotheses about Nature, and the testing of them rigorously by persistent experimentation."[11]

Recent developments in the history, sociology and philosophy of science have witnessed a radical departure from this vision of science as a system of formal propositions or discoveries, or as the mathematisation and experimental testing of hypotheses. Systematically opting for detailed case studies of the processes through which knowledge and associated skills, practices and instruments are created in preference to supra-historical grand narratives or "big-picture" accounts, scholars in these fields have in the past decades brought to light the negotiated, contingent, and situated nature of the propositions, skills, and objects that constitute natural knowledge. This scholarship has convincingly shown that scientific research is based not so much on discoveries or on logical step-by-step reasoning as on spatially and temporally situated pragmatic judgements, much like in practical crafts.[12]

In consequence, mathematics and experimental philosophy, long held to epitomise science, have now to share their once exclusive place with a host of other domains like navigational astronomy, surveying, geographical exploration, ethnography, linguistics, natural history and medicine, subjects which have been at least as significant as the former in modern scientific activity. And, although the laboratory still remains the predominant site of knowledge production for science studies, historians of science have been increasingly turning their attention to knowledge-making activities outside the strict precincts of segregated spaces, to the world of intercultural encounter, trade, politics and material exchange and their intimate historical links with the world of science.[13]

---

11   The quotations are from Needham, "Poverties and Triumphs", pp. 9 & 12, and Idem, *Science and Civilisation in China*, 7 vols. (Cambridge: Cambridge University Press, 1954–2004), vol. 7, pt. 2, p. xlvi. For a general review of the various approaches to the status of western and non-western learning traditions, see Hart (1999).

12   For a detailed presentation of these recent developments, see Golinski (1998).

13   See for example, Curtin (1984), Stewart (1999), Smith and Findlen (2002), Schiebinger and Swan (2005) and Cook (2007).

full embellishment at last

Extending this line of inquiry, Simon Schaffer has recently attended to the larger political, religious and commercial systems that enabled the accumulation of facts in early modernity. These interdependent complex networks were themselves rich in knowledge, as each of the relevant communities possessed vast amounts of intelligence, expertise and erudition about the objects – both material and immaterial – that they transacted, as well as about the context of their production and circulation. Taking as his example the sources for the data upon which Newton based his *Principia Mathematica*, Schaffer persuasively argues that these systems were decisive in the process of accumulation, extraction, evaluation, exchange and distribution of scientific knowledge. He also shows that the geography of these networks extended well beyond England or Europe: it covered almost all the continents and was coincident with that of European maritime trade and Catholic missionary movements on a global scale. For instance, Newton relied on measures garnered by sheep farmers from southwest England, at least one American mariner, and Kinh and Chinese maritime pilots from the Gulf of Tonkin, thanks to the ubiquitous English East India Company. Following Manuel Castells, Schaffer refers to these heterogeneous systems, which are composed of long- and short-range networks of trade, capital, material exchange and the movement of individuals, professional communities and religious formations, as information orders.[14]

Information orders provide a useful way here to investigate the dynamics of knowledge making during what the editors of this volume have called the Asian "Bright Dark Ages", thus allowing us to map the geography and vectors of knowledge flows in pre-1500 Asia. In what follows, this essay will thus try and show that this geography was not coterminous with civilisational spaces, but rather coincided with the spaces of commercial, political and religious circulations organised around each of the places where the knowledge in question was organised.[15] Through a comparison with the knowledge practices and information orders of early modern Europeans, it will also bring to light the deep similarities between the two. Finally, by showing that the dynamics laid out in this essay stretch until at least the end of the 18th century both in Asia and Europe, it will suggest a path that avoids the guileless Eurocentrism of the history of science, which over most of the 20th century has sought to establish the legitimacy of modern science by seeking and reifying a "scientific revolutionary" moment in the 16th and 17th centuries unique to Europe.

Jones

but where practised?

---

14   Schaffer (2009), Castells (1989) and Bayly (1996). The suggestion here is that in contrast to formalised knowledge, information describes matters more broadly shared and less explicitly challenged. Compare with Wyatt (1972).

15   On my conception of circulation and spaces of circulation, see Raj (2013).

I shall focus on two concrete examples of recognised scientific activity in Asia, which I must at the very outset confess concern places, languages and periods outside the purview of my own competence. My only reason for moving away from my own area of research here is to take up the gauntlet of rethinking Needham's Grand Question in a pan-Asian context by engaging, in the first of the two examples, with some of the very material which the great historian himself used to argue his case, in particular an early 13th-century text from southern China, the *Chu-fan-chï* (sic).[16] In the second part, I shall move to another well-documented case from another part of the Asian continent, the celebrated Greek-Arabic translation project undertaken by the Abbasids between the 8th and 10th centuries. However, unlike Needham and other Asianists, I shall not enter into the philological aspects of the knowledge in question, but shall attempt in each case to reconstitute the information, or knowledge, orders upon which these undertakings relied. I shall base myself on the work of reputed authorities, in the first case on the vast and highly detailed critical apparatus which accompanies the English translation of the *Chu-fan-chï*, and, in the second, the recent and growing scholarship that has recast this vast enterprise in the light of new research on the global history of Islamic expansion. I thus treat the Chinese example first before shifting focus to the other big scene of scientific activity in Asia which this case itself points towards, rather than following the chronological order which would have reversed the order of my narrative.

2   13th Century Song China

Let us start by examining the nature and role of the knowledge regime related to the first example, that of the *Chu-fan-chï* (literally, "Description of Barbarous Peoples") compiled in 1225, and included at the beginning of the 15th century in the *Yung-lo-ta-tien* ("Great Canon of Yung-lo"), a great collection of Chinese scientific and literary works. It appeared in print for the first time in 1783 as part of the Han-lin scholar Li-T'ian-yüan's collection of literature, known as the *Han-hai*. Edited and translated into English just over a century ago by the renowned American scholar-diplomat, William Woodville Rockhill (1854–

---

16   Chau Ju-kua, *Chau Ju-Kua: His Work on the Chinese and Arab Trade in the Twelfth and Thirteenth Centuries, Entitled Chu-fan-chï*, translated by Friedrich Hirth & William Woodville Rockhill (1911). For pragmatic reasons, mainly for coherence in citations taken from this work, I have kept the Romanised transcription of Chinese as it appears in this edition in favour of the more recent convention.

1914) and Friedrich Hirth (1845–1927), a one time maritime customs inspector in Canton and later professor of Chinese at Columbia University, it is a remarkable work on the geography, ethnography, natural history and the political and maritime trade structures of the world as known to Southern Song Chinese in the 13th century. Although dismissed in a full-page review in *The New York Times* at the time of its publication as "eloquent of [China's] self-satisfied ignorance," it has nonetheless been used as a credible source of information by present-day Sinologists.[17] The *Chu-fan-chï* has been used mainly as a source of information about China's international commercial relations during this period.[18] Others have used it to examine what the Chinese knew about foreign objects.[19] Needham too occasionally refers to the work, but again these references are cursory, in support of his assertions about the existence of Chinese seaborne trade and the knowledge and use of various techniques and technical devices, such as the compass. (Needham 1970: 42 & 245) In short – and unsurprisingly given the received, anachronistic conceptions of science as being essentially composed of disciplines such as physics, chemistry, mathematics, astronomy etc. – Chau's work has generally been seen at best as encyclopaedic, at least as simply inventorial. However, a closer reading of the text can help throw light on the intercontinental knowledge order within which it was produced.

The author of the *Chu-fan-chï*, a Song functionary named Chau Ju-kua, was an inspector of foreign trade in the province of Fujian province and probably based in the port city of T'suan-chou (now spelt Quanzhou) – one of the few facts known about him through a brief notice of his life in a descriptive catalogue of his family library.[20] In all likelihood, it was his profession that provided him the opportunity to compile this work which is composed of two parts: the first on the ethnology and political structures of the places, nations and tribes of the world known through sea-trade carried out by Chinese and Muslim traders. It is important to note here that T'suan-chou was probably the biggest seaport in the eastern hemisphere at the time and hosted a sizeable population of Arab and Persian merchants. The second part of the text focuses

---

17   "Old Chinese Book Tells of the World 800 Years Ago", *The New York Times*, December 29, 1912, Magazine Section, Part V, p. 6.

18   See for instance Abulafia (1987).

19   For example, see Digby (1982), pp. 125–159, on pp. 141–142; Donkin (2003), pp. 18, 155–161; and Claudine Salmon (2005).

20   This, as all other information cited here not directly accessible from Chau Ju-kua's text, is taken from the carefully researched, and richly documented introduction and detailed, critical footnotes to the English edition of the text by Hirth and Rockhill (1911), pp. 1–39.

on 43 articles of trade produced in these nations and of particular interest to the Chinese. The map of this world that emerges through the list of places and peoples mentioned provides an idea of the extent of the trading zone and the astonishing swathe of territory it covered. If it is no surprise to find Japan, Korea, Formosa, the Philippine Islands, Hainan, Tonkin, Annam, Kamboj, Borneo, Panrang, Java, Sumatra, the Malay Peninsula, Burma, the Andamans, Coromandel, Ceylon, Malabar, Gujarat, Malwa and India in Fujian's commercial space, this geography starts getting more remarkable when we find the Hadramaut, Oman, Baghdad, Basra, Mecca, Ghazni, Mosul, Asia Minor, Somalia, Egypt, Zanzibar and Madagascar, and simply amazing when we discover Alexandria, Sicily, southern Spain and Morocco!

The items which Chau describes in the second part of the work include olfactory and medicinal resins such as aloes, ambergris, camphor, frankincense, myrrh, dragon's blood, and asafoetida; perfumes like rose-water; dyes such as lakawood and gardenia blossoms; woods – ebony, musk-wood, sandalwood; spices – nutmegs, cloves, pepper and cardamom; fruits such as coconut, areca nuts and jackfruit; ornamental and jewellery items including ivory, tortoiseshell, opaque glass, pearls, live parrots and kingfishers' feathers; panaceas-cum-aphrodisiacs such as rhinoceros horn, and finally commonplace goods – mats and cotton. And for all these items, which are part of natural history, Chau mentions not only their origins, but also their uses in different parts of the world. Besides, as a customs official, he states the relative qualities of many of them and often gives the provenance of the most valuable ones.

These accounts of both places and produce, although brief, are fairly credible and have a ring of familiarity even to this day. Besides, in respect of the articles of commerce, Chau takes pains to differentiate between their various provenances and respective qualities. For example, on the subject of pearls he writes:

> The *chön-chu*, or "real pearls", which come from certain islands in the land of the Ta-shï [Hadramaut] are the best. They also come from the two countries of Si-nan [Ceylon] and Kién-pi [Sumatra]. Pearls are even found in [the Chinese provinces of] Kuang-si and Hu-peï, but less brilliant than those of the Ta-shï and of Kién-pi.[21]

Often, he details the way certain items are produced or extracted, sometimes even mentioning professional risks involved. To stay with the example of pearls, we read:

---

21   Chau Ju-kua, Op. cit., p. 229.

Whenever pearls are fished for they make use of thirty or forty boats, with crews of several dozens of men to each. Pearl fishers, with ropes fastened around their bodies, their ears and noses stopped with yellow wax, are let down into the water about 200 or 300 feet or more, the ropes being fastened on board. When a man makes a sign by shaking the rope, he is pulled up. Before this is done, however, a soft quilt is made as hot as possible in boiling water, in order to throw over the diver the moment he comes out, lest he should be seized with a fit of ague and die. They may fall in with huge fishes, dragons, and other sea monsters [sharks] and have their stomachs ripped open or a limb broken by collision with their dorsal fins. When the people on board notice even as much as a drop of blood on the surface of the water, this is a sign to them that the diver has been swallowed by a fish. Cases occur in which the pearl-fisher makes a signal with his rope and the man holding it on board is not able to pull him up; then the whole crew pull with all their strength, and bring him up with his feet bitten off by a monster.[22]

Given the value of pearls, especially perfectly round ones, it is not surprising to learn that they were the object of considerable traffic: "Foreign traders coming into China are in the habit of concealing pearls in the lining of their clothes and in the handles of their umbrellas, thus evading the duties payable upon them."[23] This was also the case with kingfisher feathers (t'sui-mau): "Although, of late years, the use of this luxury has been strictly forbidden by the government, the well-to-do classes still continue to add it to their dress, for which reason foreign traders, in defiance of the law, manage to smuggle it in by concealing it in the cotton lining of their clothes."[24]

Significantly, this section also details knowledge about the habitat and behaviour of kingfishers and the technique of catching them: "Kingfishers' feathers are got in great quantities in Chön-la [Cambodia], where the birds are brought forth in nests built by the side of lakes or ponds in the depths of the hills. Each pond is the home of just one male and one female bird; the intrusion

---

22 Ibid., p. 230.
23 Ibid.
24 Ibid., p. 235. On the following page, Hirth and Rockhill cite the imperial edict of 1107 banning kingfisher feathers: "The Ancient Rulers in their governmental measures extended the principle of humanity to plants, trees, birds and beasts. Now the depriving of living creatures of their life, in order to get their plumage for a perfectly frivolous purpose, is certainly unworthy of the kindness extended by the Ancient Rulers to all creatures. We therefore order the officials to stop the practice on pain of punishment."

of a third bird always ends in a duel to the death. The natives taking advantage of this peculiarity, rear decoy birds, and walk about with one sitting on the left hand raised. The birds in their nests noticing the intruder, make for the bird on the hand to fight it, quite ignoring the presence of the man, who, with his right hand, covers them with a net, and thus makes them prisoners without fail." It was this practice that had deeply shocked the Song emperors who then issued an edict in 1107 forbidding the use of kingfishers' feathers on ethical grounds, a pointer to the fact that their ethics were not just introspectively practiced at the expense of knowledge about nature, but rather was informed by the latter as much as by human interaction with nature – in foreign lands to boot!

In the case of rhinoceros horn, much in demand in China to this day, Chau gives an account of its different states: "[H]unters shoot [the rhinoceros] with a stiff arrow from a good distance, after which they remove the horn, which in this state is called a 'fresh horn' whereas, if the animal has died a natural death the horn obtained from it is called a 'dropped-in-the-hills horn'. The horn bears marks like bubbles. Those which are more white than black are the best."[25] He identifies the provenance of rhino horn as the Malay Peninsula, Tonkin, Annam, Java, India, Zanzibar and the Somali coast, from where according to him the finest horns come.

It is indeed easy to see the reason for Chau's interest in foreign people and trade structures in the fact that he was a customs official. Although he himself never left Fujian, he freely copied from older Chinese sources, particularly from Chóu K'o-feï, a high-ranking government official half a century before him who, while serving in Guangnandong (now Guangdong-Guangxi) province, collected and wrote a detailed account of China's sea trade, trade routes, the Chinese shipping and trade inspectorate, and customs duties levied on different goods. The work was known under the name of *Ling-wai-tai-ta* (literally "Notes from the Land Beyond the Passes"). However, as was customary in China at the time, and as was to be the fate of his own writings, Chau, like Chóu, did not acknowledge his sources.[26]

However, the *Chu-fan-chï* was much more than a simple compilation of earlier Chinese writings. Chau not only meticulously noted eye-witness accounts as reported to him by passing sea-men and merchants, he also had access to foreign written sources, as a lot of his information concords closely with medieval Arab texts. Thus, he lifts most or all of his information on west Asia, Asia Minor, Spain, Pemba Island and Madagascar nearly verbatim from Chóu

---

25   Ibid., p. 233.
26   Hirth & Rockhill, "Introduction", in Ibid., pp. 22 et seq.

K'o-feï's *Ling-wai-tai-ta*, or from dynastic histories, occasionally adding a few words of own. However, the sections on Sumatra, Malay Peninsula, Java, Ceylon, Malabar, Gujarat, Malwa, Zanzibar, Yemen, Somali coast, Oman, Kish Island, Basra, Egypt, Alexandria, Andaman Islands, the fabulous city of Djabulsa, Sicily, Morocco, Philippine Islands and Formosa seem to be based exclusively on oral information from Chinese and foreign traders. Although the text itself does not mention any of these interlocutors, one can get some idea of the nature of the knowledge economy of the transcontinental merchant networks of the time through some of the documents from the Cairo Geniza, notably those studied by S.D. Goitein for the western Indian Ocean. Indeed, these throw valuable light on the process of knowledge circulation in this vast maritime world about a staggering variety of subjects ranging from the properties and relative qualities of goods and the fluctuation of their prices to political, religious and social happenings in different parts of this space.[27]

In some chapters, such as those on Annam, Panrang, Cambodia, Burma, the Coromandel Coast and Ghazni, the author augments traders' accounts that he records directly with paragraphs from Chóu. The entry on Hainan is taken mainly from dynastic histories, but it also contains a lot of original material.[28] And occasionally, he transposes passages which had been recorded under one region in earlier texts, especially the *Ling-wai-tai-ta*, to another region, because he had learnt from other sources that that was where they really belonged. Thus, while Chóu described the ostrich in his section on Madagascar, Chau faithfully copied it ... in his chapter on the Berbera (Somali) coast, where the bird properly belongs. Indeed, in their introduction to the work and the exhaustive notes to each entry, Hirth and Rockhill meticulously analyse the origins of each of Chau's assertions.

It is worth quoting an extract from the *Chu-fan-chï* to give an idea of the kind and credibility of information recorded in it. Thus, on Egypt we find:

> The people live on cakes, and flesh; they eat no rice. Dry weather usually prevails. The government extends over sixteen provinces, with a circumference of over sixty stages. When rain falls the people's farming ... is washed out and destroyed. There is a river (in this country) of very clear and sweet water, and the source whence springs this river is not known. If there is a year of drought, the rivers of all other countries get low, this river alone remains as usual, with abundance of water for farming purposes, and the people avail themselves of it in their agriculture. Each

---

27 Goitein (1973, 2008) Ghosh (1992). See also Reinaud (1845).
28 Hirth & Rockhill (1911), pp. 36–37.

succeeding year it is thus, and men of seventy or eighty years of age cannot recollect that it has rained . . . Furthermore there is a city called Kié-yé [Cairo] on the bank of this river . . . In this river there are water-camels (cranes?) and water-horses [hippopotami] which come up on the bank to eat the herbs, but they go back into the water as soon as they see a man.[29]

Chau then goes on to describe the lighthouse at Alexandria:

The country of O-kön-t'o [Alexandria] belongs to Wu-ssï-li [Egypt]. According to tradition, in olden times a stranger . . . built on the shore of the sea a great tower under which the earth was dug out and two rooms are made, well connected and very well secreted. In one vault was grain, in the other were arms. The tower was two hundred *chang* [feet] high. Four horses abreast could ascend to two-thirds of its height. In the centre of the building was a great well connecting with the big river. To protect it from surprise by troops of other lands the whole country guarded this tower that warded off the foes. In the upper and lower parts of it twenty thousand men could readily be stationed to guard, or to sally forth to fight. On the summit there was a wondrous great mirror; if war-ships of other countries made a sudden attack, the mirror detected them before-hand, and the troops were ready in time for duty. In recent years there came (to O-kön-t'o) a foreigner, who asked to be given work in the guard-house of the tower; he was employed to sprinkle and sweep. For years no one entertained any suspicion of him, when suddenly one day he found an opportunity to steal the mirror and throw it into the sea, after which he made off.[30]

Not surprisingly, the *Chu-fan-chï* also contains descriptions of fabulous places, some mythical, like "the countries of women" supposedly situated somewhere in the Malay archipelago: "The women of this country conceive by expos-ing themselves naked to the full force of the south wind, and so give birth to female children."[31] Others, no less fabulous but not necessarily apocryphal,

---

29   *Chu-fan-chï*, pp. 144–145.
30   Ibid., p. 146.
31   Ibid., p. 151. This passage is again taken from the *Ling-wai-tai-ta*, but seems to be fairly widely known; for example Antonio Pigafetta, the Venetian chronicler of Magellan's cir-cumnavigation of 1519–1522, reports having heard of "an island called Acoloro, which lies below Java Major, [where] there are found no persons but women, and that they become

include Ssï-kia-liyé (Sicily) – "this country has a mountain with a cavern of great depth in it; when seen from afar it is smoke in the morning and fire in the evening; when seen at a short distance it is a madly roaring fire." – and Ch'a-pi-sha (Djabulsa, the land of the setting sun) – "this country is resplendent with light, for it is the place where the sun goes down. In the evening when the sun sets, the sound of it is infinitely more terrifying than that of thunder, so every day a thousand men are placed at the gates who, as the sun goes down, mingle with the sound of the (sinking) sun that of the blowing of horns and the beating of gongs and drums. If they did not do this, the women with child would hear the sound of the sun and would die of fright."[32]

Apart from the descriptions of places, people and some of their notable customs, Chau at times also attempts to situate places or territories with respect to each other. For instance, in the section on the Coromandel coast, or the Chola dominion (Chu-lién), he states that it is in the peninsular part of India, its capital being "five *li* distant from the sea; to the west one comes to Western India [after] 1500 *li*; to the south one comes to [Ceylon after 2500 *li*]; to the [east] one comes to the [Malay peninsula after 3000] *li*." Or, when describing Baghdad (Pai-ta), "the great metropolis of all the countries of the Arabs," he states that it is reached "by land from Oman . . . after about 130 days journey, passing on the way some fifty cities;" and Ghazni (Ki-tz'ï-ni) "from Ma-lo-pa [Oman] in about a hundred and twenty stages. The country lies to the north-west [of India], and is exceptionally cold, the winter's snow not melting until the spring." And, if one travels north from Spain (Mu-lan-pi) for two hundred days, "the days are only six hours long."[33]

Thus, the descriptions in the *Chu-fan-chï* of polities (especially when discussing the Islamic world), extent of territories, Sino-Arab relations, routes, goods produced, measures, etc., are the result of complex and heterogeneous circulations of older Chinese texts, the embodied knowledge of Chinese and foreign (mainly Arab and Persian, but certainly also Indian, Malay and Mediterranean) merchants and sailors as well as written sources in Arabic and, undoubtedly, other Asian languages. This work on world geography, ethnography, maritime trade structures, and natural history is the outcome of forms of knowledge made in very different contexts and times by diverse practitioners who had little in common, all of which were brought together by Chau in his work. In short, the work is the outcome of short- and long-range circulations of

---

pregnant from the wind." See his *Magellan's Voyage around the World*, tr. J.A. Robertson, bi-lingual (Italian-English) text, 3 vols. (1906), v. 2, pp. 169–171.

32   *Chu-fan-chï*, pp. 153–154.

33   Ibid., pp. 93–94; 135; 138; 143 respectively.

persons, goods, skills, knowledge, etc., extending in the past and present, from China right up to Spain and Morocco, the western limits of the Islamic world – its information order covering the very same space that it describes.

It is clear from the above that the *Chu-fan-chï* was not a one-off text produced by an idiosyncratic Song bureaucrat, but can be situated within a wider corpus of similar works to which it is clearly related.[34] Much less was it the result of some autarkical epistemology peculiar to the Chinese, a civilisational trait supposedly based on Confucianism, an inward-looking doctrine obsessed with human affairs and the right conduct that had no room for the wider world and nature.[35] Rather, it demonstrates that the dynamic of Chinese culture itself depended on sustained trans-regional interactions covering large swathes of the world known both directly and indirectly to 13th-century Chinese, especially interactions with the world of west-Asian trade. It thus points to the inextricable historical links between knowledge-making, trade, bureaucratic and fiscal practices, links which historians of modern science have only recently begun to examine, albeit in the context of "*western*" science, but which clearly underlie knowledge practices in other societies and periods as well.[36]

Now, since Needham equated Asia with China and alluded to other India or West Asia only to declare that the religious and state impediments to scientific development he identified in China could be extended to these spaces, it is interesting to take a brief look at least one other region to assess the validity of his thesis in the broader Asian context. Early Islamic Asia provides an ideal terrain inasmuch as both religion and state are crucial components of social organisation.

3   The Abbasid Empire, 8th–10th Centuries

Let us then turn to another period in history: the early Abbasid empire and the celebrated translation project, from the second half of the 8th century, which lasted until the end of the 10th century, of classical Greek learning into Arabic started by the Abbasids in their capital city Baghdad. As is widely acknowledged, it was thanks to this gigantic undertaking that almost all non-literary and non-historical secular Greek works, i.e., in philosophy and science, notably in mathematics, logic, astronomy, astrology, alchemy, medicine

---

34   For the historical tradition in China and wider context of such works, see Park (2012), especially Ch. 1. See also Heng (2009); and Yangwen (2011).

35   Needham, *scc*, vol. 2, p. 12.

36   On this point, see, for instance, Salguero (2010–11).

and mechanics, that were available throughout the eastern Byzantine empire and west Asia were translated into Arabic. (Gutas 1998: 1) It was these Arabic translations that served as the basis for their survival and translation into Latin and thus for their later appearance in Christian Europe. In fact, many Greek texts would never have been known had they not been saved in translation through Arabic, the originals having over time been definitively lost. (Toomer 1984) While a number of European academics have tended to minimise the role of Islam by portraying it as a passive transmitter of knowledge, this effort has been held up by others as incontrovertible proof of the scientific temper of Islam and its enlightened rulers, and ultimately of their crucial contribution to the development of Science and Human Progress. The story is, however, more complicated than either narrative suggests.[37]

The translation process was in fact inaugurated for a number of reasons. For one, the Arab conquests from the 7th century onwards had brought vast swathes of territory across three continents, from Central Asia and the Indian subcontinent to North Africa and Spain, under a single polity, rendering possible the free flow of goods, raw materials, people, techniques, skills, ideas and modes of thought. Thus, not only did this "Pax Islamica" oversee an unprecedented impetus to trade, so too did it witness massive plant migration between Asia, Europe and North Africa as well as the introduction and spread of paper and paper-making techniques across the empire in the wake of the Abbasid conquest of the western fringes of Chinese central Asia.[38] The need to meet the practical challenges introduced in the light of these new multifarious circulatory phenomena gave rise to the translation, appropriation and further elaboration of the ancient knowledge heritage of the region. The domains included agriculture, construction of technological devices for irrigation and commerce, astrological divination, medicine, administration of justice in a multi-confessional empire and problems of surveying, to name but a few.

In addition to these pragmatic considerations, the translation movement as a sustained long-term undertaking was also in large measure a consequence of the imperial ideology of the early Abbasids seeking to legitimise themselves as the true successors of the Sasanian emperors in the eyes of the Persian communities who were instrumental in their victory over their Umayyad predecessors. Indeed, the Sasanians had already been engaged in translating Greek and Sanskrit texts into Pahlavi as they believed that these works were

---

37   For an excellent analysis of the actual process of translation and its implications for the communication of scientific thought, albeit in the much more recent setting of the late 19th century, see Elshakry (2008).

38   See Watson (1983); and Bloom (2001).

themselves derived from the Avesta, a corpus which in their eyes formed the totality of applied and theoretical knowledge and which had been pillaged and scattered in the wake of Alexander's invasion of Persia in myriad translations. This process continued even after the collapse of the Zoroastrian Empire, not least to educate the bureaucracy, and was adopted by the first caliph al-Mansur (r. 754–775) and his son al-Mahdi (r. 775–785) in their effort to win over the region's literati.[39] The *Bayt al-Hikma* (House of Wisdom) library founded by the Abbasids in Baghdad, a successor to an earlier and modelled on the fabled library in Alexandria, further testifies to the consciousness of the strategic role of knowledge in the sustained existence and consolidation of their rule.

This vast, multiform and, for a large part, fortuitous process of circulation and translation of knowledge thus mobilised members of almost all the literate communities within this Islamic "commonwealth", which included Muslims, Chalcedonian, Nestorian and Monophysite Christians, Copts, Sabaeans, Jews, Avestans, Buddhists and Hindus. Thus, merchants and landowners, caliphs and princes, architects and builders, astrologers and astronomers, clerics, perfumers, medics, alchemists, geometers, tax collectors, military commanders and other state functionaries, etc., of religious, cultural, ethnic and linguistic origins as diverse as the above-mentioned material and immaterial entities they were involved in these transactions.

In this context, it then becomes clear that the Greek tradition was not the only source for the new Arabic science. For, Greek, albeit the most important source language, was not the only one in this translation exercise – texts in Syriac, Pahlavi and Sanskrit were also part of this gigantic project, sometimes even as sources for Greek originals. For instance, Cassianus Bassus's *Eklogai peri georgias* ("Selections on farming") was initially translated around the 7th century from an earlier Pahlavi translation and only later from the Greek. (Ullmann 1972: 434–435) Another example is that of Theophilus of Edessa (d. 785), a Christian and military astrologer at caliph al-Mahdi's court, who was widely familiar with Greek, Syriac, Pahlavi and Indian sources, and a broad range of Indian and Persian texts were even available in Syriac translation by Nestorian missionaries established in the East.[40] Thus, multilingual professionals like Theophilus, of whom there were many in the empire, could draw on living knowledge traditions in languages other than Greek and communicate with each other directly through travel or correspondence. Also, governors

---

39   Gutas (1998), Ch. 5. See also Lecomte, "L'introduction du Kitāb Adab al-Kātib d'Ibn Qutayba," *Mélanges Louis Massignon*, 3 vols. (Damascus: Institut français de Damas, 1956–57), v. 3, pp. 45–64.

40   Gutas (1998), p. 16. See also Conrad (1995).

of Abbasid-controlled provinces, such as Sind, could, and did, send cultural-scientific missions to Baghdad, mainly in the late 8th or early 9th century, when the control was strongest. One such mission involved an Indian emissary from Sind to Baghdad in 772–73, "an expert in calculations [of] the motions of the planets" [who] had with him in a book [*siddhanta*] consisting of 12 chapters, equations made according to *kardaja*'s [arguments] calculated for each half degree, together with various operations on the sphere for the two eclipses, the ascensions of the signs and other things."[41] This work, which has been identified as an astronomical text on the theory of star cycles (*kalpas*), composed in Sanskrit in Rajasthan in the early 7th century, was translated into Arabic by Muhammad ibn Ibrahim al-Fazari (d. c.800) as the famous *Zij al-Sindhind* and played a crucial role in the development of early Abbasid astronomy. However, this was not the first infusion of Indian astronomical theories into Islam, as a number of texts had already been translated in the Persian realm since at least the early 8th century.[42]

It is important to remark here that unlike Greek, which, for want of practitioners especially in mathematics and astronomy, was well-nigh dead by the 6th century as a language of specialised learning, the other major languages of the empire were still being used for formulating and communicating learned thought. This implied, amongst other things, that the translation of certain specialist Greek texts required not only a translator but also a corrector, as the former did not always understand the contents of what they were translating, leading to ambiguities or plain errors. Ptolemy's *Almagest* and Euclid's *Elements* were amongst the first works to be translated from the Greek. The quality of the translation was often faulty and Greek texts had to be translated more than once, the *Elements* being translated twice and the *Almagest* at least three times. (Hogendijk 1996) In a recent article, the historian of science, Mohammed Abattouy and his colleagues contend that this semantic operation even required the active participation of specialists and scholars from outside the Greco-Arabic world, in particular from Persia and India, to help

---

41   al-Fazārī (1970), pp. 105–106. See also al-Biruni, tr. E.C. Sachau (1888), v. 2, p. 15; and Pingree
        (1968).
42   On the subject of Indian and Iranian astronomy and astrology, see Pingree (1963) and Paul
        Kunitzsch (2003). For a carefully researched account of the social relations between the
        Arab conquerors and the local populations of Sind, see Maclean (1989). I am particularly
        grateful to Derry Maclean for generously sharing some of his vast knowledge of these rela-
        tions to help me make my way through the secondary literature on which this section is
        based. It goes without saying that he bears no responsibility in any erroneous interpreta-
        tions this account might contain.

reconstruct the contents of certain Greek texts. According to them, the result was then not a simple translation, but a "transformation of scientific knowledge in an intercultural context, a process without which a mere transmission of texts would be meaningless."[43] Without prejudging the veracity of their claim, it is at least highly plausible that the translations from all these diverse traditions were not kept isolated from each other in a kind of massive repository of learning, but were used and eclectically mixed by practitioners who in all likelihood confronted these knowledge forms in various practical contexts.

In this example, we once more witness decentralised, heterogeneous networks of material and immaterial entities – principally people, goods, embodied knowledge, books, know-how, states and religion – in play in the making of this other "Asiatic" knowledge economy. As with the *Chu-fan-chï*, its constituent knowledge order was not coterminous with the Islamic world, but maps onto the wider world beyond its precincts – for instance, Zoroastrian, Hindu, Jewish and Christian knowledge practices, Chinese, Sasanian, and Hindu polities and religious orders, not to mention the multi-confessional merchant communities upon whose networks much of the knowledge flows depended. Although there are obvious similarities in the inextricable relationship between the worlds of trade, bureaucracy and knowledge between the making of the *Chu-fan-chï* and this huge translation project, the major difference between the two is the significant presence in the latter case of the state in patronising the making, appropriation, capitalisation and dissemination of knowledge. Indeed, it is not until the 19th century that one can discern signs of these forms of state patronage and intervention in the sciences. At any rate, both examples portray an image of knowledge making at the antipodes of a stagnant "Asiatic mode of production" mired in a despotic bureaucratic feudalism busy expropriating and recycling surpluses from peasant subjects into defence and public works, and preventing the emergence of capitalism and science.[44]

On the contrary, Asian bureaucracies were far more versatile than Needham's conception would allow for. In the Chinese case, they were also concerned with trade and thus eager to know about the wider world; in the Abbasid empire, they were preoccupied as much with human improvement as with questions of law, measurement, accurate balances and practical mathematics. In short, the Asian ethos resembled that of early modern Europe more than Weberian comparativism would have us believe. In particular, all the characteristics presented here show a close resemblance between Asian and early-modern and modern "European" scientific activity, not to speak of a similarity in the con-

---

43   See also Abattouy, Renn & Weinig (2001).
44   Needham, *scc*, v. 7, pt. 2, p. 5.

But what then happened?

tent of the knowledge produced. It is to this comparison that we shall in conclusion now turn.

4 Enter Europe – Exit Asia?

Indeed, so similar were the general practices and the knowledge regimes on which they were based that in the wake of European expansion into the Indian Ocean in the early 16th century, Europeans experienced little difficulty in interacting with the already existing spaces of knowledge circulation in the region. This is particularly visible in the case of Garcia da Orta (c. 1500–1568) whose *Coloquios dos simples e drogas e cousas medicinais da India* ("Colloquies on the Simples, Drugs and other Medicinal Substances of India") published in Goa in 1563 is considered a foundational text for early modern botany. Orta was a prominent Portuguese medic who, at the time of publishing his book, had spent 30 years in India. Setting sail from Lisbon for Goa in 1534 as the personal physician of Martim Affonso de Sousa, the Captain-Major of Portuguese Asia, he spent the rest of his life in and around the Portuguese settlements on the subcontinent's west coast, practicing medicine both in Goa and in various Deccan courts, notably that of Ahmadnagar.[45] In addition to enjoying a lease on the island of Bombay, where he cultivated a garden of medicinal plants, he invested the considerable wealth he acquired from his lucrative medical practice in shipping and trade – chiefly in materia medica and precious stones. As such, Orta interacted closely with Persian, Malay, Arab and Indian traders, apothecaries and physicians and also maintained a network of paid correspondents and agents for the supply of plants, seeds and other merchandise from across the Indian Ocean region. His reputation, though, rests mainly on the *Colloquies*, or dialogues: the first lay printed publication from Goa – and South Asia – which continues to this day to be considered authoritative on many of the subjects of which it treats.

The *Coloquios* deals mainly with Asian pharmacopoeias and the botanical knowledge of the region from where these pharmacopoeias originated. Like the trade on which part of Orta's livelihood and fortune depended, this text too is based on the same heterogeneous knowledge-rich networks of specialised trade, politics, religion and medicine as we witnessed in the earlier Asian cases. Organised in the form of a dialogue between the author and an apocryphal Spanish colleague who visits Orta in Goa, the work contains 57 chapters

---

45 The biographical details of Garcia da Orta are based on Carvalho (1934). See also Boxer (1963).

on drugs and simples, mostly of vegetable origin but also ivory, diamonds and bezoar, with the names of each substance in the different languages of the regions where it was to be found, or traded, a description and discussion of its country or region of origin, the commercial routes by which it reached Goa, its medicinal and therapeutic uses and the parts used. On reading the book it is clear that Orta depended on the same intermediaries for his Asian botanical and medical knowledge as he did for his trade and political relations.

This work, which also seriously called into question prevalent medical beliefs in Europe at the time by confronting them with South Asian theories, was to enjoy great renown in Europe through its Latin translation and was to become a fundamental text for natural historical instruction, notably in Leiden. A comparison of the *Coloquios* with the *Chu-fan-chï* shows that they both have very similar structures and the quality and form of knowledge is very similar, although the latter is far more cursory in its treatment of the articles it describes. It must be said, however, that the *Chu-fan-chï* did not claim to be a pharmacopoeia but presented the knowledge of a customs official. On the other hand, it is incontrovertible that both texts fed on the interaction and confluence of similar variegated communities and were thus based on comparable knowledge orders, and were in this sense akin to the knowledge order that undergirded Newton's work as described by Simon Schaffer.

During the course of the 17th and 18th centuries, an increasing number of Europeans, both lay and religious, riding on the expanding infrastructures of European trading empires successfully entered the well-established knowledge-rich regimes of the Indian Ocean world on which knowledge making in the region intrinsically depended. It is crucial to note here that the presence of these newcomers did not imply the displacement of the indigenous knowledge communities, but as in the case of Garcia da Orta, worked in close association with, and piggy-backed on, the latter. (Cook 2007) The resulting knowledge was thus in no way a purely European produce as I have shown elsewhere. It involved the crucial input of a vast number of Asian networks, yet it formed a fundamental and conspicuous part of the knowledge that was being accumulated and capitalised in Europe. (Raj 2007) Indeed, as Simon Schaffer has shown, even those European savants who did not step outside their countries, let alone Europe, relied for their own projects on distant, often Asian but not exclusively, knowledge-rich networks as well as knowledge produced through them. (Schaffer 2009)

Although, like Needham, then, this essay also argues against an *ex nihilo* rise of science in Europe, the history outlined here does not lean towards a vision of scientific activity on the world stage as a game of musical chairs,

but performance ???

where one occupant necessarily displaces another at different times in the game. Science was, has been, and still is, an interconnected activity with many centres of accumulation. Its story cannot be told in terms of divergence. In this sense, it did not ever disappear from Asia to take root in Europe instead. Indeed, Asia comprised a vital part of the making of scientific knowledge even where Europe was involved. However, it was differentially appropriated and accumulated in each part, much as has been persuasively been argued by David Washbrook for the closely related question of Modernity. (Washbrook 1997) It is certainly true that between the 1830s and 1914 Asia, significant parts of which were either under European domination or else in severe crisis, were side-lined on the global economic and scientific scene, but already parts of it were reappearing on front stage by the end of the First World War.

This chapter then suggests that beyond differences of locality, culture, economy, period, the role of religion and the state, etc., there are certain features which all these knowledge-making activities from East and West share and which cannot be reduced to the conventional division between Western and non-Western science. These features form significant enabling conditions for the flourishing of modes of knowledge production, circulation and translation in, and between, a variety of fields, both more "practical" and more "theoretical" in their production and use, which should not be separated and dealt with as if unrelated to each other. More broadly, they also hint at a revision of the view of the emergence of modern science and its periodisation, but also of present-day sciences and knowledge in general.

In so doing, this chapter reiterates its allegiance to the transdisciplinary tradition of recent science studies which while maintaining one foot firmly anchored in historian's history, seeks also to engage with classic questions of philosophy and sociology of science from a interdisciplinary perspective. It does, though, take its distance from the deeply entrenched Eurocentrism of most writings of my fellow science studies scholars.

In sum, notwithstanding that two trees do not make a forest, and that a lot more work needs to be done than my own competence allows, this essay hopefully indicates a possible route towards the replacement of the outmoded and inherently Eurocentric concept of civilisation and an ahistorical idea of science by a more relevant conceptual vocabulary to help us understand science as a phenomenon of a global humanity with specific local and institutional variations. It is also a nod in the direction of transregional research in the history and political economy of the sciences and their inextricable relationship to trade, religion and the state, an approach which treats each set of actors on their own terms.

Bibliography

Abattouy, Mohammed, Jürgen Renn & Paul Weinig. "Transmission as Transformation: The Translation Movements in the Medieval East and West in a Comparative Perspective," *Science in Context* (2001) 14: 1–12.

Abulafia, David (1987). "Asia, Africa and the Trade of Medieval Europe," in Michael M. Postan and Edward Miller (eds.), *The Cambridge Economic History of Europe*, Vol. 2: *Trade and Industry in the Middle Ages*, 2nd ed. Cambridge: Cambridge University Press, pp. 402–473.

al-Fazārī, Mohammad. *Fragment Z1*, translated by David Pingree, "The Fragments of the Works of al-Fazārī", *Journal of Near Eastern Studies* (1970) 29: 103–123.

al-Biruni, Muhammad ibn Ahmad. *Alberuni's India: An Account of the Religion, Philosophy, Literature, Geography, Chronology, Astronomy, Customs, Laws and Astrology of India.* Translated by E.C. Sachau (1888), 2 vols. London: Trübner & Co.

Bayly, C.A. (1996). *Empire and Information: Intelligence Gathering and Social Communication in India, 1780–1870.* Cambridge: Cambridge University Press.

Bloom, Jonathan M. (2001). *Paper before Print: The History and Impact of Paper in the Islamic World.* New Haven, CT: Yale University Press.

Boxer, C.R. (1963). *Two Pioneers of Tropical Medicine: Garcia d'Orta and Nicolás Monardes.* London: The Hispanic Luso-Brazilian Councils.

Butterfield, Herbert (1949). *The Origins of Modern Science.* London: G. Bell & Sons.

Castells, Manuel (1989). *The Informational City: Information Technology, Economic Restructuring and the Urban Regional Process.* Oxford: Basil Blackwell.

Carvalho, Augusto da Silva (1934). *Garcia d'Orta: Comemoração do quarto centenário da sua partida para a India em 12 de março de 1534.* Coimbra: Imprensa da Universidade.

Chau Ju-kua (1911). *Chau Ju-Kua: His Work on the Chinese and Arab Trade in the Twelfth and Thirteenth Centuries, Entitled Chu-fan-chï,* tr. Friedrich Hirth & William Woodville Rockhill, with critical notes and annotations. St. Petersburg: Imperial Academy of Sciences.

Conrad, Lawrence I. (1995). "The Arab-Islamic Medical Tradition," in Lawrence I. Conrad, Michael Neve, Vivian Nutton, Roy Porter & Andrew Wear (eds.), *The Western Medical Tradition: 800BC–1800AD.* Cambridge: Cambridge University Press, pp. 93–138.

Cook, Harold J. (2007). *Matters of Exchange: Commerce, Medicine, and Science in the Dutch Golden Age.* New Haven, CT: Yale University Press.

Cunningham, Andrew & Perry Williams. "De-centring the 'Big Picture': The Origins of Modern Science and the Modern Origins of Science," *British Journal for the History of Science* (1993) 26: 407–432.

Curtin, Philip D. (1984). *Cross-Cultural Trade in World History.* Cambridge: Cambridge University Press.

Digby, Simon (1982). "The Maritime Trade of India," in Tapan Raychaudhuri & Irfan Habib, (eds.), *The Cambridge Economic History of India*, Vol. 2. Cambridge: Cambridge University Press, pp. 125–159.

Donkin, Robin A. (2003). *Between East and West: The Moluccas and the Traffic in Spices Up to the Arrival of Europeans*. Philadelphia, PA: American Philosophical Society.

Duhem, Pierre (1913–1959). *Le système du monde*, 10 vols. Paris: A. Hermann.

Elshakry, Marwa S. "Knowledge in Motion: The Cultural Politics of Modern Science Translations in Arabic," *Isis* (2008) 99: 701–730.

Elvin, Mark. "Why China Failed to Create an Endogenous Industrial Capitalism: A Critique of Max Weber's Explanation," *Theory and Society* (1984) 13: 379–391.

Ghosh, Amitav. "The Slave of MS. H.6," in Partha Chatterjee & Gyanendra Pandey (eds.), *Subaltern Studies. Writings on South Asian History and Society*, Vol. 7. New Delhi: Oxford University Press, pp. 159–220.

Godelier, Maurice (1964). *La notion de "mode de production asiatique" et les schémas marxistes d'évolution des societies*. Paris: Centre d'études et de recherches marxistes.

Goitein, Shelomo Dov (1973). *Letters of Medieval Jewish Traders*. Princeton: Princeton University Press.

——— (2008). *India Traders of the Middle Ages: Documents from the Cairo Geniza*. Leiden: E.J. Brill.

Golinski, Jan (1998). *Making Natural Knowledge: Constructivism and the History of Science*. Cambridge: Cambridge University Press.

Gutas, Dimitri (1998). *Greek Thought, Arabic Culture: The Graeco-Arabic Translation Movement in Baghdad and Early 'Abbasid Society (2nd–4th/8th–10th Centuries)*. London: Routledge.

Habib, S. Irfan & Dhruv Raina (eds.) (1999). *Situating the History of Science: Dialogues with Joseph Needham*. New Delhi: Oxford University Press.

Hall, A. Rupert (1954). *The Scientific Revolution 1500–1800: The Formation of the Modern Scientific Attitude*. London: Longmans, Green & Co.

——— (1983). *The Revolution in Science, 1500–1750*. Harlow: Longman.

Hamilton, Gary. "Why No Capitalism in China? Negative Questions in Historical Comparative Research," *Journal of Developing Societies* (1985) 1: 187–211.

Hart, Roger. "Beyond Science and Civilization: A Post-Needham Critique," *East Asian Science, Technology, and Medicine*, (1999) 16: 88–114.

Heng, Derek (2009). *Sino-Malay Trade and Diplomacy from the Tenth through the Fourteenth Century*. Athens, OH: Ohio University Press.

Hogendijk, Jan P. (1996). "Transmission, Transformation, and Originality: The Relation of Arabic to Greek Geometry," in F. Jamil Ragep & Sally P. Ragep, with Steven Livesey (eds.), *Tradition, Transmission, Transformation: Proceedings of Two Conferences on Pre-Modern Science Held at the University of Oklahoma*. Leiden: E.J. Brill, pp. 31–64.

Koyré, Alexandre (1958). *From the Closed World to the Infinite Universe*. New York: Harper.

———— (1968). *Metaphysics and Measurement: Essays in Scientific Revolution*. Cambridge, MA: Harvard University Press.

Kunitzsch, Paul (2003). "The Transmission of Hindu-Arabic Numerals Reconsidered", in Jan P. Hogendijk & Abdelhamid. I. Sabra (eds.), *The Enterprise of Science in Islam: New Perspectives*. Cambridge, MA: MIT Press, pp. 3–21.

Lecomte, Gérard. "L'introduction du Kitāb Adab al-Kātib d'Ibn Qutayba," *Mélanges Louis Massignon*, 3 vols. (Damascus: Institut français de Damas, 1956–57), v. 3, pp. 45–64.

Maclean, Derryl N. (1989). *Religion and Society in Arab Sind*. Leiden: E.J. Brill.

Mazlish, Bruce (2004). *Civilization and Its Contents*. Stanford, CA: Stanford University Press.

Needham, Joseph (1954–2005). *Science and Civilisation in China*, 7 volumes. Cambridge: Cambridge University Press.

———— (1969). *The Grand Titration. Science and Society in East and West*. London: George Allen & Unwin.

———— (1970). *Clerks and Craftsmen in China and the West: Lectures and Addresses on the History of Science and Technology*. Cambridge: Cambridge University Press.

Parthasarathy, Prasannan (2011). *Why Europe Grew Rich and Asia Did Not: Global Economic Divergence, 1600–1850*. Cambridge: Cambridge University Press.

Park, Hyunhee (2012). *Mapping the Chinese and Islamic Worlds: Cross-Cultural Exchange in Pre-Modern Asia*. Cambridge: Cambridge University Press.

Pigafetta, Antonio. *Magellan's Voyage around the World*, tr. J.A. Robertson, bi-lingual (Italian-English) text, 3 vols. (1906). Cleveland, OH: The Arthur H. Clark Co.

Pingree, David. "The Fragments of the Works of Ya'qub ibn Tariq," *Journal of Near Eastern Studies* (1968) 27: 97–125.

———— "Astronomy and Astrology in India and Iran," *Isis* (1963) 54: 229–246.

Pomeranz, Kenneth (2000). *The Great Divergence: China, Europe and the Making of the Modern Economy*. Princeton: Princeton University Press.

Raj, Kapil (2007). *Relocating Modern Science: Circulation and the Construction of Knowledge in South Asia and Europe, 1650–1900*. Basingstoke: Palgrave Macmillan.

———— "Beyond Postcolonialism . . . and Postpositivism: Circulation and the Global History of Science," *Isis* (2013) 104: 337–347.

Reinaud, Joseph Toussaint. *Relation des voyages faits par les Arabes et les Persans dans l'Inde et à la Chine dans le IXe siècle de l'ère chrétienne*. Texte arabe imprimé en 1811 par les soins de feu Langlès publié avec des corrections et additions et accompagné d'une traduction française et d'éclaircissements, 2 vols. Paris: Imprimerie royale, 1845.

Rosenthal, Jean-Laurent and R. Bin Wong (2011). *Before and Beyond Divergence: The Politics of Economic Change in China and Europe*. Cambridge, MA: Harvard University Press.

Salguero, C. Pierce. "Mixing Metaphors: Translating the Indian Medical Doctrine *Tridosa* in Chinese Buddhist Sources," *Asian Medicine* (2010–11) 6: 55–74.

Salmon, Claudine (2005). "La diffusion du gong en Insulinde vue essentiellement à travers divers épaves orientales (période Song-Ming)," in Jorge M. dos Santos Alves, Claude Guillot & Roderich Ptak (eds.), *Mirabilia Asiatica: Produtos raros no comércio maritime*. Wiesbaden: Otto Harrassowitz, pp. 89–116.

Schaffer, Simon. "Newton on the Beach: The Information Order of the *Principia Mathematica*," *History of Science* (2009) 47: 243–276.

Schiebinger, Londa and Claudia Swan (eds.) (2005). *Colonial Botany: Science, Commerce, and Politics in the Early Modern World*. Philadelphia, PA: University of Pennsylvania Press.

Sivin, Nathan (1995). *Science in Ancient China: Researches and Reflections*. Aldershot: Ashgate.

———— "Why the Scientific Revolution Did Not Take Place in China – or Didn't It?" *Chinese Science* (1982) 5: 45–66.

———— (1985). "Max Weber, Joseph Needham, Benjamin Nelson: The Question of Chinese Science," in Eugene Victor Walter, Vitautus Kavolis, Edmund Leitis & Marie Coleman Nelson (eds.), *Civilizations East and West: A Memorial Volume for Benjamin Nelson*. Atlantic Highlands, N.J.: Humanities Press, pp. 37–50.

———— (1983). "Chinesische Wissenschaft: Ein Vergleich der Ansätze von Max Weber und Joseph Needham," in Wolfgang Schluchter (ed.), *Max Webers Studie über Konfuzianismus und Taoismus. Interpretation und Kritik*. Frankfurt: Suhrkamp, pp. 342–362.

Smith, Pamela H. and Paula Findlen (eds.) (2002). *Merchants and Marvels: Commerce, Science and Art in Early Modern Europe*. New York & London: Routledge.

Stewart, Larry. "Other Centres of Calculation, or, Where the Royal Society Didn't Count: Commerce, Coffee-Houses and Natural Philosophy in Early Modern London," *British Journal for the History of Science* (1999) 32: 133–153.

Subrahmanyam, Sanjay (2001). "Inde ouverte ou Inde fermée?" in Yves Michaud (ed.), *Qu'est-ce la culture?* Vol. 6. Paris: Odile Jacob, pp. 69–79.

———— (2013). *Is Indian Civilization a Myth? Fictions and Histories*. Ranikhet: Permanent Black.

Toomer, Gerald J. "Lost Greek Mathematical Works in Arabic Translation," *The Mathematical Intelligencer* (1984) 6: 32–38.

Thapar, Romila (1999). "History of Science and the *Oikoumene*," in Habib & Raina (1999), pp. 16–28.

Ullmann, Manfred (1972). *Die Natur- und Geheimwissenschaften im Islam*. Leiden: E.J. Brill.

Washbrook, David. "From Comparative Sociology to Global History: Britain and India in the Prehistory of Modernity," *Journal of Economic and Social History of the Orient* (1997) 40: 410–443.

Watson, Andrew M. (1983). *Agricultural Innovation in the Early Islamic World: The Diffusion of Crops and Farming Techniques, 700–1100*. Cambridge: Cambridge University Press.

Weber, Max. "Konfuzianismus und Taoismus", *Archiv für Sozialforschung*, 41 (1915), translated into English by Hans H. Gerth as *The Religion of China: Confucianism and Taoism* (Glencoe, IL: The Free Press, 1951).

Werskey, Gary (1988). *The Visible College: A Collective Biography of British Scientists and Socialists in the 1930's*. London: Free Association Books.

Wong, R. Bin (1997). *China Transformed: Historical Change and the Limits of European Experience*. Ithaca, NY: Cornell University Press.

Westfall, Robert S. (1992). *The Scientific Revolution in the 17th Century: The Construction of a New World View*. Oxford: Clarendon Press.

Wittfogel, Karl A. (1931). *Wirtschaft und Gesellschaft Chinas; Versuch der wissenschaftlichen Analyse einer grossen asiatischen Agrargesellschaft*. Leipzig: C.L. Hirschfeld.

Wyatt, H.V. "When does Information become Knowledge?" *Nature* (14 Jan. 1972) 235: 86–89.

Yangwen, Zheng (2011). *China on the Sea: How the Maritime World Shaped Modern China*. Leiden: E.J. Brill.

Zilsel, Edgar (2003). *The Social Origins of Modern Science*. Dordrecht: Kluwer Academic Publishers.

Index of Names

Index of Subjects

No ref to Collins
Predictability?

22 Main approaches, Whigs etc.
23. points of fusion. Needham
24. Not one ocean. key
43 Key PS in western
59. bourgeois ignored
60 Corinna? + 62
62 Core defn
65. Kuhn's defn of science
77 Key. Independent science
78. rare use of context
+79   ''
     Sigurdsson excellent ch
91 useful conclusion
97 Arab influence on science
121 Indian numerals key
          Contribution
+126
129 Bacon. how. why
131 Axioms as new
134 European additions
141 Baconian power
144 maps as learning
163. Study of networks
     not the jamu per se

173. Special nature of medicine
     (Key to applicability
        question)
     Focus on practice, not
        representation
187. 0 + cultural + linguistic
        filters.

251 Key comparisons
  | cause Key
     the debate
253

205. big row. Too narrow
210 science defined by Needham
212. Chr. science defined
224. Why bother!!
227. Loss of one's preferred cultural
          symbolism

228 Muslim science instrumental
229 but saved its soul
230 Key summary of debate
236 key. multicausality
252 big caveat/conclusion?
256 Needham summary ✓
     Why then ignore it across the book

259 Embeddedness proper
268 role of trade OK but! 1225
269 Islam collectors Abbasid
275 key